Online Scheduling in Manufacturing

T0191712

Haruhiko Suwa · Hiroaki Sandoh

Online Scheduling in Manufacturing

A Cumulative Delay Approach

 Springer

Haruhiko Suwa
Department of Mechanical Engineering
Setsunan University
Neyagawa, Osaka
Japan

Hiroaki Sandoh
Graduate School of Economics
Osaka University
Toyonaka, Osaka
Japan

ISBN 978-1-4471-5827-1   ISBN 978-1-4471-4561-5 (eBook)
DOI 10.1007/978-1-4471-4561-5
Springer London Heidelberg New York Dordrecht

© Springer-Verlag London 2013
Softcover reprint of the hardcover 1st edition 2013
This work is subject to copyright. All rights are reserved by the Publisher, whether the whole or part of
the material is concerned, specifically the rights of translation, reprinting, reuse of illustrations,
recitation, broadcasting, reproduction on microfilms or in any other physical way, and transmission or
information storage and retrieval, electronic adaptation, computer software, or by similar or dissimilar
methodology now known or hereafter developed. Exempted from this legal reservation are brief
excerpts in connection with reviews or scholarly analysis or material supplied specifically for the
purpose of being entered and executed on a computer system, for exclusive use by the purchaser of the
work. Duplication of this publication or parts thereof is permitted only under the provisions of
the Copyright Law of the Publisher's location, in its current version, and permission for use must always
be obtained from Springer. Permissions for use may be obtained through RightsLink at the Copyright
Clearance Center. Violations are liable to prosecution under the respective Copyright Law.
The use of general descriptive names, registered names, trademarks, service marks, etc. in this
publication does not imply, even in the absence of a specific statement, that such names are exempt
from the relevant protective laws and regulations and therefore free for general use.
While the advice and information in this book are believed to be true and accurate at the date of
publication, neither the authors nor the editors nor the publisher can accept any legal responsibility for
any errors or omissions that may be made. The publisher makes no warranty, express or implied, with
respect to the material contained herein.

Printed on acid-free paper

Springer is part of Springer Science+Business Media (www.springer.com)

Preface

This book provides results and findings from our research in online scheduling. Online scheduling is recognized as the crucial decision-making process of production control at a phase of "being in production" according to the released shop floor schedule. Online scheduling can also be considered as one of key enablers to realize prompt *capable-to-promise* as well as *available-to-promise* to customers along with reducing production lead times under recent globalized competitive markets.

The main part of this book is to introduce new approaches to online scheduling based on a concept of *cumulative delay*. The cumulative delay is regarded as consolidated information of uncertainties under a dynamic environment in manufacturing and can be collected constantly without much efforts at any points in time during a schedule execution. In our approach, the cumulative delay of the schedule has the important role of a criterion for making a decision whether or not a schedule revision is carried out. We believe that the cumulative delay approach to trigger schedule revisions has the following capabilities for the practical decision making:

1. To reduce frequent schedule revisions which do not necessarily improve a current situation with much expense for its operation;
2. To avoid overreacting to disturbances dependent on strongly an individual shop floor circumstance; and
3. To simplify the monitoring process of a schedule status.

This book will be of interest to both practitioners and researchers who work in planning and scheduling in manufacturing. Researchers will find the importance of when-to-revise policies during a schedule execution and their influences on scheduling results which have been obtained by numerous computer simulations. Practitioners will benefit from the broad range of topics in scheduling from theoretical issues to applications. We hope all readers of this book will achieve new insights into techniques for scheduling.

   This book is partially supported by the Grants-in-Aid for Scientific Research of
the Japan Society for the Promotion of Science. Finally, we are deeply grateful to
our friends, colleagues, and our family, in particular Anthony Doyle, Simon Rees,
Claire Protherough, Grace Quinn, and Christine Velarde, who helped us with the
completion of the book.

Osaka, Japan, March 2012   Haruhiko Suwa
                                                                             Hiroaki Sandoh

Contents

Part I
Introduction

Chapter 1
Introduction

1.1 Background

Production scheduling is essential in manufacturing to enable cost effective and timely production, and thereby to meet the deadlines of individual products with target qualities. In the actual environments, jobs are initially assigned to the limited number of resources over a specified time horizon, which will be measured by a shift or a day in many cases. The sequences of the jobs to be processed on the individual resources are then determined under some suitable criterion associated with at least one performance measure. The schedule obtained in this manner is released to the shop floor as referential information to control the production activities and to manage material procurement, shipments of orders, and the other external activities as well.

The rapid progress of Information and Communication Technology (ICT) in these two decades has made information systems one of key enablers to realize flexible and adaptive manufacturing systems [30]. Basically information technology is a collective term of information processing, information communication and accumulation of information to share. Information-sharing, among others, has led to realization of Concurrent Engineering (CE) which emphasizes the quick response to customers [14, 36]. Concurrent engineering, which is occasionally referred to as simultaneous engineering [12, 17], is a significant paradigm and/or concept in product development involving several aspects of product design and planning, and has directly influenced upon scheduling strategies.

In fact, there have been growing requirements in manufacturing for developing new production control systems which can concurrently handle decision-makings associated with scheduling on the shop floor and higher level decision-making such as capacity planning [10, 19]. There exist scheduling systems which incorporate process planning and/or production planning. Hegde et al. [18], for example, has reported a case study on the integrated scheduling systems with simultaneous production planning for carbon products. A computer integrated system has also been developed, which can provide the system users with the feedback information from process planning and scheduling [2]. For large scale manufacturing systems, a concurrent decision-making system has been developed which can cope with process planning and scheduling concurrently based on computational simulation [40]. They implemented the decision-making system into a virtual factory [15] or a virtual production system [16] which simulates generalized factories in the actual environments, and demonstrated its effectiveness.

Moreover, developments in Supply Chain Management (SCM) have been expected to realize the global optimization ranging from material procurement to sell-out in recent competitive global markets [4, 7, 21, 32]. This is because global optimization can reduce the related cost more efficiently than the combination of local optimization, and because the recent development in the ICT has enabled us to carry out such global optimization more easily than before.

Weintraub et al. [40] have developed a problem-solving system which integrates the scheduling techniques with the higher level decision-makings associated with process planning, production planning and Material Requirement Planning. Advanced Planning and Scheduling (APS) and a new concept of scheduling so called Drum-Buffer-Rope (DBR) also deserve our attention. In the above, the APS is a kind of production control system [22], which intends to realize global optimization in supply chain by concurrently processing from available-to-promise to progress management on the shop floor through material requirement programming, scheduling and operation instructions. The DBR, on the other hand, is one of the underlying ideas in Theory of Constraints (TOC) by Goldratt [13] to realize efficient manufacturing [5, 41]. It is an abbreviation for three terms; the drum which figuratively represents the capacity of a bottleneck stage, the buffer which is a slack duration, and the rope indicating the timing of production progress.

As we have observed in the above, scheduling has significantly adapted itself to the ongoing manufacturing technologies as well as drastic changes in social and economic conditions and technical innovation. This implies that scheduling in manufacturing is a crucial decision-making process in the production control phase. Some practitioners refer to scheduling activities, in a metaphorical sense, as "the control stick" of a large and complicated manufacturing system, and therefore, scheduling decisions might be likened to controlling the stick skillfully according to rapid changes of circumstances.

1.2 Introduction to Online Scheduling

Figure 1.1 describes a general information flow under manufacturing circumstances. In Fig. 1.1, short-term capacity planning, for relatively short time horizon such as a month is categorized into the decision-making process to allocate resources and to specify worker shifts along with a shop floor schedule, according to the results of the upper-level capacity planning based on a master production schedule and customer orders. Short-term capacity planning is usually carried out by utilizing a Capacity Requirement Planning (CRP) system within a Material Requirement Planning (MRP) system in modern manufacturing. Then the shop floor schedule, in other words, the

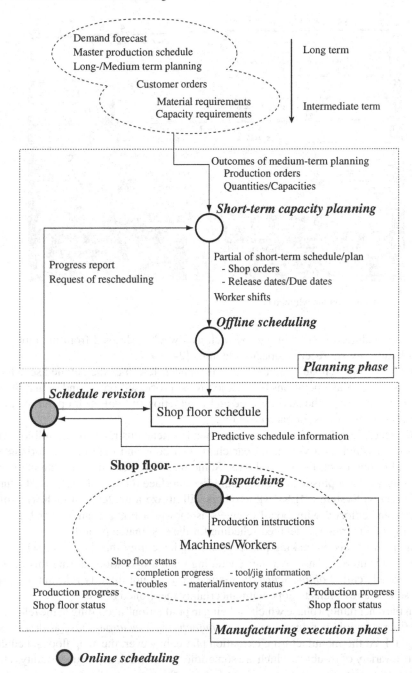

Fig. 1.1 Information flow focused on schedules

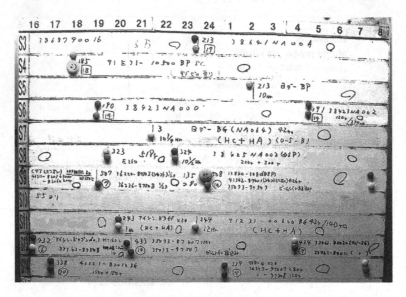

Fig. 1.2 Gantt chart on the whiteboard

detailed schedule for a certain period such as a week is derived from the monthly plan obtained by short-term capacity planning [24, 39].

The shop floor schedule obtained in this manner describes the detailed schedule information such as operations of each job to be processed on the individual manufacturing resources, the starting time and completion time of each operation, and production instructions for each resource.

The schedule released into the shop floor is often illustrated by a well-known Gantt chart which is a visualized bar chart with hour and date on the horizontal axis and resources on the vertical axis. Some factories make use of computerized schedules with a sophisticated Graphical User Interface (GUI) under a Manufacturing Execution System (MES), while others still utilize a whiteboard with wipe-off markers and colorful whiteboard magnets as shown in Fig. 1.2, and/or stick notes skillfully to illustrate the detailed schedule for the designated period.

The above decision making phase before implementing the shop floor schedule is called a planning phase. All the scheduling related operations during this phase is generically called *offline scheduling*, since they are not directly relevant to shop floor schedule revisions to cope with uncertainty mentioned below.

On the other hand, a phase which is "being in production" according to the released shop floor schedule is referred to as a *manufacturing execution phase* as described in Fig. 1.1. At the manufacturing execution phase, however, the shop floor schedule suffers a variety of troubles, which are sometimes substantial, due to uncertainty. The sources of uncertainty may be machine failures, fluctuations of setup times caused by, e.g., inexperienced workmen, and the other possible troubles on the shop floor. The sources of uncertainty may be also urgent orders or changes of due dates which are directed by the higher level.

Uncertainties, which are not expected or are not taken into account at the planning phase, will possibly, cause delays on the shop floor schedule, that is, the actual shop floor schedule deviates from the planned one. Accumulation of these delays will significantly disturb the smooth progress of the whole schedule and may result in low productivity of the manufacturing system and the violation of due date promising. In some situations, the schedule may lose its feasibility when the total delay time becomes very large.

As depicted in Fig. 1.1, some modifications to or some adjustments in the existing shop floor schedule become necessary to adapt these delays in the manufacturing environments. This is generically called *online scheduling*. The "online" signifies the state of "being in production" of a manufacturing system where the environments dynamically change from moment to moment due to the influences of internal and external activities. Online scheduling can be interpreted and recognized in various ways:

- Online scheduling is the dynamical process to monitor and control the progress of shop floor schedules.
- Online scheduling is, in addition, a real-time decision making process to cope with uncertainties such as disruptions and unforeseen events happening in real time.

Online scheduling involves the decision makings called *dispatching* and *schedule revision*. The terminology "dispatching rule" is, in scheduling theory, used to indicate a rule which regulates the order of jobs to be relegated to their individually assigned resources. In this book, however, "dispatching" is also used to express giving detailed production instructions to resources on the shop floor, and is occasionally used to represent the production instructions themselves.

Schedule revision is a main part of online scheduling, which is the ongoing decision making to modify or to adjust the current shop floor schedules appropriately according to changes in the manufacturing environments in order to realize the smooth progress of manufacturing. When the major disruptions or unexpected events occur which might cause serious delay of the schedule, the decision-maker must decide whether or not to request rescheduling for the higher level decision making prior to schedule revision, where rescheduling signifies to create a new shop floor schedule from scratch after canceling the unprocessed jobs from the current schedule.

In some ways we can identify online scheduling as much more crucial decision making that does not allow optimism than offline scheduling.

1.3 Key Issues in Online Scheduling

Online scheduling is to entail greater risks than offline scheduling with regard to time restrictions for schedule coordination, stability of the shop floor status, and internal and external costs incurred by re-allocation of resources and re-sequence of operations. However, we can occasionally come across the real circumstances where online scheduling or quasi-online scheduling is carried out manually and skillfully, even though comprehensive methodologies for online scheduling have not yet been established.

The major issues in consolidating concepts and methodologies for online scheduling are:

- Management of uncertainty,
- The timings of decision makings relevant to online scheduling,
- Severe restrictions that all the decision makings must be carried out promptly in real time.

These issues are considered in the remainder of this section. Among others, the emphasis is placed on how to manage uncertainty and the timings of the scheduling decisions.

Management of Uncertainty

In building online scheduling model, one of the obstacles to be removed is a problem how well they can manage uncertainties. A representative approach will be an event-driven one reacting to contingencies directly, though an execution of scheduling under an event-driven approach will correspond to a schedule revision under the online-scheduling in this book.

Disadvantages of the event-driven approach are discussed as follows: There is a risk of incurring a huge cost for frequent schedule revisions under an event-driven approach, and therefore some suitable classification of events will be required according to the effectiveness of the schedule revisions since every event has a possibility of triggering a schedule revision wasting a huge cost as well as time.

Events can be classified into critical events and non-critical ones [25]. Typical examples of critical events are a machine breakdown requiring half a day to restore and processing of an urgent job with a strict due date promising, and that of non-critical events is a slight fluctuation of a setup time for manual tool change by an untrained worker. Even if we confine ourselves to the above mentioned critical events only, non-critical events might occasionally become identical to a critical events when they are accumulated. These observations reveal that some adequate management method to cope with uncertainty is required for online scheduling.

Timing of Decision-Making

As a certain perspective, it can be claimed that decision-makings of online scheduling contains the followings [29]:

- **How-to-revise**:

  - Where on the shop floor schedule the scheduler should revise; The whole shop floor schedule or the partial of the current shop floor schedule.
  - How the schedule should be revised.

  It should be noted in the above that if the whole shop floor schedule is to be revised, it will be sent to the higher level decision making.

- **When-to-revise**:

  - When the decision-maker should make a judgement on whether or not a schedule revision should be carried out.
  - When the scheduler should revise the current shop floor schedule.

  A number of studies on online scheduling seems to have addressed to the above how-to-schedule problem, that is, they emphasize methods for schedule revisions by means of knowledge-based technology [23, 26, 31] and through constraints satisfaction modeling [3, 11, 33] on the basis of an event-driven approach. On the other hand, studies on the when-to-revision problems can seldom be seen.

  As for the timing of a schedule revision, the following three types are realistic [20, 37, 38]:

- **Periodic type**: Schedule revisions are to be carried out periodically, where the time between tow consecutive schedule revisions are pre-determined.

- **Event-driven type**: The floor schedule is revised when a some new event occurs.

- **Hybrid type**: The shop floor schedule is revised at the prespecified periodic planned time and when an event occurs.

- **Enhanced event-driven type**: The shop floor schedule is revised when an event occurs or when the elapsed time since the most recent revision reaches the planned time, whichever occurs first.

These four policies are discussed in details in Chap. 3.

Restrictions in Real Time

In real circumstances, once the manufacturing system starts its operations, it is crucial not to suspend the manufacturing system during its operations. Even if critical events such as machine breakdowns and interruptions by urgent jobs should occur, it is important to cope with these events as promptly as possible. In this case, information relevant to these events should be collected without delay and aggregated consistently with each other. For this purpose, we need a some suitable mechanism and/or production system as depicted in Fig. 1.3 to continually collect and aggregate

Fig. 1.3 Production information system

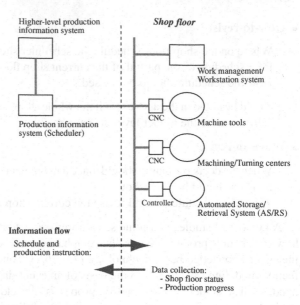

information relevant to every kind of event which might suspend the manufacturing system. Online scheduling in this book aims to consider such a system or mechanism which can perform in a restricted time.

Another Issues in Online Scheduling

Online scheduling problems have a variety of factors that are relevant to each other complicatedly, for example, the progress of the shop floor schedule and the amount of its delay, the times when individual uncertain events occur and the sizes of their influences upon the progress, and so on, all of which depend on the manufacturing environments and system scale. This indicates that it is very difficult to deal with online scheduling problems systematically. Only a few studies have attempted to discuss online scheduling as well as semi-online scheduling on the conditions that all the information associated with inputs such as the number of jobs and their processing times are unknown or that some of the information are, a priori, known [1, 8, 9, 27], which are to be considered in Chap. 2. The some topics of online scheduling problems can be found in [6, 28, 34, 35]

1.4 Outline of Book

This book is composed by three parts. The first part includes Chap. 2 and this chapter. In Chap. 2, the framework of scheduling methods is overviewed, and then scheduling as a decision-making is clarified using some classifications such as *deterministic* scheduling versus *stochastic* scheduling, *static* scheduling versus *dynamic* scheduling, and/or *offline* scheduling versus *online* scheduling. A wide variety of scheduling problems are also summarized based on manufacturing environments, restrictions and constraints, and the performance measures.

Part II consisting of Chaps. 3–5 deals with models of online scheduling. In the real circumstances, a variety of uncertainties often inhibits the smooth progress of a planned schedule. Therefore it becomes an essential activity to release quickly a new schedule after revising the current schedule adaptively against the uncertainties. But when and how should a schedule be revised?—Chap. 3 deals with timings of schedule revisions and Chap. 4 describes methods for online scheduling. In Chap. 3, the *periodic* schedule revision policy and the *event-driven* schedule revision policy are introduced as basics of the timing. The advanced policies are also described. They are called the *hybrid* schedule revision policy which has the advantages of both the periodic and the event driven policy, and the *enhanced* schedule revision policy. Chapter 4 introduces the basic procedures of dispatching rules and then describes the ways of schedule revision. In this chapter, knowledge-based approaches, which have played important roles in online scheduling, are also introduced.

Chapter 5 introduces a new approach to online scheduling, in which the *cumulative delays* of the schedule is employed as a measure to make a judgement whether or not a schedule is revised. Chapter 5 also describes the concept and definition of cumulative delays and then discusses schedule revision policies based on the cumulative delays. In those policies, schedule revision is invoked when the current cumulative delay exceeds a prescribed threshold called a *critical cumulative delay* denoted by D^*. By incorporating the concept of monitoring cumulative delays into the schedule revision policies, the three new policies, which are called a D^*-*driven policy*, a *hybrid D^*-driven policy* and an *enhanced D^*-driven policy*, are developed.

Part III consisting Chaps. 6–8 deals with applications of cumulative delay based schedule revision policies to some shop models. In Chap. 6, the D^*-driven policy is applied to dynamic job shops with random machine breakdowns. Chapter 7 focuses on the application of the hybrid D^*-driven policy to single machine configurations with urgent jobs and spontaneous delays of processing times. In Chap. 8, the effectiveness of the enhanced D^*-driven policy is demonstrated by applying it to dynamic flexible flow shops with urgent jobs.

References

1. Albers S (2010) Online scheduling. In: Robert Y, Vivien F (eds) Introduction to scheduling. Chapman and Hall/CRC, Boca Raton, pp 51–73
2. Aldakhilallah KA, Ramesh R (1999) Computer integrated process planning and scheduling (CIPPS): intelligent support for product design, process planning and control. Int J Prod Res 37(3):481–500
3. Baptiste P, Pape CL, Nuijten W (2001) Constraint-based scheduling—applying constraint programming to scheduling problems. Kluwer Academic Publishers, Norwell

4. Brewer AK, Button KJ, Hensher DA (eds) (2001) Handbook of logistics and supply-chain management. Elsevier Science, Oxford
5. Chakravorty SS, Atwater JB (2005) The impact of free goods on the performance of drum-buffer-rope scheduling systems, Int J Prod Econ 95(3):347–357
6. Chrétienne P, Coffman EG Jr, Lenstra JK, Liu Z (1995) Scheduling theory and its applications. Wiley, Chichester
7. Christopher M (1998) Logistics and supply chain management—strategies for reducing cost and improving service, 2nd edn. Pearson Education, London
8. Cottet F, Delacroix J, Kaiser C, Mammeri Z (2002) Scheduling in real-time systems. Wiley, Chichester
9. Dandamudi S (2003) Hierarchical scheduling in parallel and cluster systems. Kluwer Academic Publishers, New York
10. Das BP, Rickard JG, Sha N, Macchietto S (2000) An investigation on integration of aggregate production planning, master production scheduling and short-term production scheduling of batch process operations through a common data model. Comput Chem Eng 24(2–7): 1625–1631
11. Elleby P, Fargher HE, Addis TR (1988) Reactive constraint-based job-shop scheduling. In: Oliff MD (ed) Expert systems and intelligent manufacturing. North-Holland, New York, pp 1–10
12. Eversheim W, Bochtler W, Gräßler R, Kölscheid W (1997) Simultaneous engineering approach to an integrated design and process planning. Eur J Oper Res 100(2):327–337
13. Goldratt EM (1997) Critical chain. The North River Press, Great Barrington
14. Haddad CJ (1994) Concurrent engineering and the role of labor in production development, Control Eng Pract 2(4):689–696
15. Halevi G (2001) Handbook of production management methods. Butterworth-Heinemann, Oxford
16. Hatono I, Nishiyama T, Umano M, Tamura H (1998) Performance evaluation of distributed real-time scheduling systems using distributed production system simulator. In: Okino N, Tamura H, Fujii S (eds) Advances in production management systems—perspectives and future challenges. Chapman and Hall, London, pp 423–434
17. Hauck WC, Bansal APE, Hauck AJ (1997) Simultaneous engineering—correlates of success. Int J Prod Econ 52(1–2)15,83–90
18. Hegde GG, Kalathur S, Tadikamalla PR, Maurer J, Abraham KP (1998) Production scheduling on parallel machines: a case study. Omega, Int J Manage Sci 26(1):63–71
19. Heijnen P, Bouwmans I, Verwater-Lukszo Z (2005) Improving short-term planning by incorporating scheduling consequences. Comput Aided Chem Eng 20:997–1002
20. Herrmann JW (2006) Rescheduling strategies, policies, and methods. In: Herrmann JW (ed) Handbook of production scheduling. Springer, New York, pp 135–148
21. Hopp WJ, Spearman ML (2001) Factory physics. McGraw-Hill Higher Education, New York, pp 582–623
22. Hvolby H-H, Steger-Jensen K (2010) Technical and industrial issues of advanced planning and scheduling (APS) systems. Comput Ind 61(9):845–851
23. Kerr R, Szelke E (1995) Artificial intelligence in reactive scheduling. Chapman and Hall, London
24. McCarthy SW, Barber KD (1990) Medium to short term finite capacity scheduling—a planning methodology for capacity constrained workshops. Eng Costs Prod Econ 19(1–3):189–199
25. Mehta SV, Uzsoy RM (1998) Predictable scheduling of a job shop subject to breakdowns. IEEE Trans Robot Autom 14(3):365–378
26. Noronha SJ, Sarma VVS (1991) Knowledge-based approaches for scheduling problems: a survey. IEEE Trans Knowl Data Eng 3(2):160–171
27. Pruhs K, Sgall J, Torng E (2004) Online scheduling. In: Leung JY-T (ed) Handbook of scheduling—algorithms, models, and performance analysis. Chapman and Hall/CRC, Boca Raton, pp 15-1-15–39
28. Robert Y, Vivien F (2010) Introduction to scheduling. Chapman and Hall/CRC, Boca Raton

29. Sabuncuoglu I, Kizililsik OB (2003) Reactive scheduling in a dynamic and stochastic FMS environment. Int J Prod Res 41(17):4211–4231
30. Shaw MJ (1998) Introduction to the special issue on information-based manufacturing. Int J Flex Manuf Syst 10:195–196
31. Smith SF (1995) Reactive scheduling systems. In: Brown DE, Scherer WT (eds) Intelligent scheduling systems. Kluwer Academic Publishers, Boston, pp 155–192
32. Statdler H, Kilger C (eds) (2000) Supply chain management and advanced planning. Springer, Heidelberg
33. Suh MS, Lee A, Lee YJ, Ko YK (1998) Evaluation of ordering strategies for constraint satisfaction reactive scheduling. Decis Support Syst 22:187–197
34. Tanaev VS, Gordon VS, Shafransky YM (1994) Scheduling theory—single-stage systems. Kluwer Academic Publishers, Dordrecht
35. Tanaev VS, Sotskov YN, Strusevich VA (1994) Scheduling theory—multi-stage systems. Kluwer Academic Publishers, Dordrecht
36. Thomas M Jr (1996) Concurrent engineering: supporting subsystems. Comput Ind Eng 31(3)–4:571–575
37. Vieira GE, Herrmann JW, Lin E (2000) Analytical models to predict the performance of a single-machine system under periodic and event- driven rescheduling strategies. Int J Prod Res 38(8):1899–1915
38. Vieria GE, Herrmann JW, Lin E (2003) Rescheduling manufacturing systems: a framework of strategies, policies and methods. J Sched 6:39–62
39. Wall B, Higgins P, Browne J, Lyons G (1992) A prototype system for short-term supply planning. Comput Ind 19(1):1–19
40. Weintraub A, Cormier D, Hodgson T, King R, Wilson J, Zozom A (1999) Scheduling with alternatives: a link between process planning and scheduling. IIE Trans 31(11):1093–1102
41. Wu SY, Morris JS, Gordon TM (1994) A simulation analysis of the effectiveness of drum-buffer-rope scheduling in furniture manufacturing, Comput Ind Eng 26(4):757–764

Chapter 2
Outline of Scheduling

Abstract At the first, the present chapter overviews the framework of a wide variety of scheduling methods, and thereby we can expect to become able to grasp the trends and perspectives in the researches on scheduling. Secondly, we observe the classification of models related to scheduling decisions. The well known classification may be *deterministic* scheduling versus *stochastic* scheduling, or *static* scheduling versus *dynamic* scheduling. In addition, we also consider a classification by *online* scheduling versus *offline* scheduling to clarify how we can discuss the online scheduling within the framework of scheduling problems. Finally, we summarize various scheduling models based on the famous triplet $\alpha|\beta|\gamma$, i.e. manufacturing environments(α), restrictions and constraints (β), and the performance measures (γ).

2.1 Overview of Frameworks and Methods

Theoretical Approaches

The study of scheduling in the academic fields started in 1950s by researchers in industrial engineering and operations research from a wide range of perspectives. The first step is Johnson's epoch-making work [62] followed by the work of Jackson [56, 57]. The work by Johnson has provided the framework of deterministic scheduling models to allocate resources to planned jobs so as to minimize one ore two performance measure such as, maximum job tardiness or total flow times. It would be fair to say that the so called Johnson rule for flow shop scheduling problems with two or three machines is the origin of the "scheduling theory" in manufacturing.

In the late 1960s, researchers in computer engineering also began to study scheduling problems with a view to handling scarce computational resources efficiently. Their findings have developed into a new research field of processor scheduling or task scheduling for hard real-time systems or online systems. The processor/task scheduling problems in computing can basically be discussed within the framework

H. Suwa and H. Sandoh, *Online Scheduling in Manufacturing*,
DOI: 10.1007/978-1-4471-4561-5_2, © Springer-Verlag London 2013

of the scheduling theory in manufacturing (e.g., a single machine scheduling problem with minimizing total completion time). The term "online" in processor or task scheduling means, in a broad sense, a state of information systems that are connected to a network or another computer or controlled by them.

In the 1970s, theoretical studies on scheduling gradually directed issues on complexity of scheduling problems [12, 24]. This is because it was shown that most of scheduling problems belong to the *NP*-hard class, indicating that we cannot reach an optimal solution even for relatively simple problems such as a single machine scheduling problem. However, many researchers attempted to devise various kinds of deterministic scheduling models with their efficient optimization algorithms. These considerable efforts have contributed to the systematization of the scheduling theory [19, 45, 80].

At the same time, *stochastic scheduling* started considering that real circumstances in manufacturing are subject to various sources of uncertainty which shall bring randomness into scheduling decisions [38, 113, 116, 129]. The sources of uncertainty, which is often confined to unforeseen events, may include fluctuations of processing times of jobs, machine breakdowns which may arise from tool breakages, releases of unscheduled orders. Most of studies addressing stochastic scheduling have attempted to analyze simple priority rules to dispatch jobs with relatively limited conditions that values of one or more attributes associated with jobs are random, e.g, release dates, processing times or due dates of jobs.

The above approaches for scheduling have mainly been discussed within the theoretical framework with a wide variety and perspectives. They have concentrated on theoretical aspects for the purpose of obtaining optimal solutions along with their complexity issues, especially for deterministic models. They have apparently been relevant to combinatorial (discrete) optimization, and have provided valuable insights toward the development of theoretical scheduling models. It is, however, to be noted that they are limited to an ideal manufacturing environment with completely accurate scheduling information.

Computing-Based Approaches

In the 1980s, researchers addressing scheduling problems from the practical point of view have focused on the development of effective priority rules, usually called dispatching rules, to dispatch jobs by means of computer simulations, aiming at local optimization or sub-optimization at an individual manufacturing stage or for an individual facility. The dispatching rules are flexible and useful tools since we can catch up with changes in a manufacturing system promptly by switching a dispatching rule to another suitable one according to the environmental changes. These studies have accumulated huge quantities of insights for useful characteristics of priority rules in the practical environments, and have led to the development of the so called *simulation-based scheduling*. The approach based on computer simulation is considered as the proper extension of the application of a single dispatching rule. The simulation-based technique have developed into the framework of reactive scheduling.

In the 1980s, furthermore, emphasized were much more practical sides of problems than the above as well as theoretical completeness of researches. A large number of studies on scheduling problems were proposed for, e.g., scheduling problems in flexible shops [35, 73] that were recognized as the extension of shop models. Problems in an flexible manufacturing system (FMS) [58, 60, 123, 145] and just-in-time production (JIT) [50, 69, 118] were also explored. The essence of FMS scheduling is to determine the routings of the jobs together with the job sequences on machines [1, 91], and most of studies on FMS scheduling used simulation approaches or heuristic ones although the target problems were described and formulated as an optimization problem in many cases [10, 123, 127].

It should be noted, in the end of the 1980s, new approaches from new other academic fields, which could be positioned between theoretical study and practical one, appeared coupled with the rapid progress of information technologies. The technologies used in these new approaches are referred to as Artificial Intelligence (AI) and Computational Intelligence (CI) [65, 102, 104]. An expert scheduling system is one of the realizations of AI-based scheduling.

The major characteristics of the expert scheduling system are that the system generates or modifies a schedule according to the knowledge stored in its database called knowledge-base [111]. The system possibly utilizes a so called inference engine in order to extract or arrange the suitable rules to the schedule from the knowledge-base. The knowledge-base is generally constructed in such a way that knowledge engineers collect useful information regarding scheduling decisions and their results, mainly through interviewing expert human schedulers, and then convert them into a logical form such as an if-then rule. This process is generically called knowledge acquisition.

However, there still exists a serious problem in the construction of the knowledge-base; all of the knowledge to be extracted from the expert scheduler are generally based on their experiences. For this reason, they are not necessarily explicit enough to describe and are not accountable in many cases. This problem has been called a *bottleneck of knowledge acquisition* for the construction of the knowledge-base.

These controversial issues have been pointed out by Wiers [146], but several successful studies have also been reported. For example, De et al. [32] have developed a knowledge-based scheduling system for a testing process in a semiconductor industry, which has been implemented into an actual manufacturing system. Their target problem was a so called generalized job shop problem, and they utilized the beam search method based on their knowledge-base. Another example is the scheduling support system developed by Huang and Lin [53], which aims at interaction with a human scheduler. Their system supports human decision-making so as to handle complicated production information efficiently and to alleviate the burden on the scheduler.

The computational intelligence approaches contain neural networks, fuzzy, evolutionary computations and so called metaheuristics [99]. Among others, a collection of metaheuristics approaches are successful because of their general-purpose properties and ease of implementation. In scheduling, metaheuristics itself means a framework to provide simple procedures for obtaining a solution to a target scheduling problem. Its representatives are:

- Genetic algorithm (e.g. [18, 28, 29, 84, 96]),
- Tabu search (e.g. [2, 17, 103, 110, 128, 143]),
- Simulated annealing (e.g. [54, 70, 79]),
- Particle swarm optimization (e.g. [9, 64, 89, 151]),
- Greedy randomized adaptive search (e.g. [3, 40–42]), and
- Ant colony optimization (e.g. [20, 122, 147]).

Resende et al. [119] have presented valuable insights of modern metaheurisics not only in scheduling problems but in other combinatorial optimization problems [25, 92, 121].

It should also be noted that over the last two decades, many researchers in operations research and industrial engineering have attempted to refine the traditional mathematical optimization methods such as Lagrangean relaxation methods [63, 93, 94] and the branch and bound methods [49, 108, 109, 125]. Lagrangean relaxation methods have attracted many attentions from researchers due to their resulting high quality schedules with relatively short computation times and their predictability of the optimality of solutions and CPU loads even though a sophisticated mathematical formulation with decompositional property [31, 46] is required.

In summary, it would be fair to say that there has being increase in research efforts toward bridging the gap between theory and practice, which has been pointed out time and again.

2.2 Classification

A wide variety of models scheduling problems can be classified according to their nature. The most well-known classification is

(A) Deterministic versus stochastic

This is a traditional classification frequently observed in industrial engineering and operations research. This classification is carried out depending on the three major properties; manufacturing environments, job properties and objective functions.

A scheduling model is called *deterministic* if all the attributes needed for construction of a schedule take constant values and they are known in advance, that is, the number of jobs and the number of machines available are, a priori, fixed and known, and the values of other attributes associated with jobs and machines also constant and known. In general, the deterministic scheduling problems are discussed within the framework of combinatorial optimization problems in operations research, where research emphases tend to be placed on the efficiency or optimality of a scheduling method and its complexity issue.

On the other hand, we can recognize that the manufacturing environments changes dynamically due to uncertainty. The factors of uncertainty contain rush orders bringing an unexpected release date of the corresponding job, tool breakages causing machine breakdowns and the other unforeseen events. From this point of view, one can model a scheduling problem taking account of these unforeseen events in a stochastic manner. When one or more attributes are expressed by random variables, the scheduling model is called *stochastic*. It is postulated under a stochastic model that the probability distribution of job attributes such as processing times and occasionally release dates are all known at the beginning of scheduling. Unlike deterministic approach, studies on stochastic scheduling models seems to have directed an analysis of the structure of specific scheduling problems rather than the development of scheduling algorithms.

It should be noted that classification (A) can be utilized only for well-defined scheduling problems. A great deal of research on scheduling has addressed the definition of a new scheduling problem and the development of its efficient scheduling algorithm(s), based on the framework of deterministic or stochastic scheduling problems. It would be no doubt that this classification has become a fundamental framework of scheduling theory because of its systematic description. It would be also noteworthy to refer the work by Pinedo [115, 116] which coherently gives detailed descriptions the classification mentioned above, the methods and their analytical results.

Static versus Dynamic

The another type of classification can be done by:

(B) Static versus dynamic

which can also be observed in the various literature. One may recognize that this type of classification is similar to the above (A). However, these classifications (A) and (B) are essentially different because classification (B) emphasizes circumstances where scheduling decisions are actually made, rather than the properties of jobs and manufacturing environments as seen in classification (A). For this reason, classification (B) does not refer to the definitions of scheduling problems at all.

A reasonable and conventional interpretation for static/dynamic will be:

- A scheduling problem is *static* if the release dates of jobs to be processed are assumed to be known in advance.
- A scheduling problem is *dynamic* if the release dates of jobs to be processed have a stochastic nature.

Static scheduling decisions do not take account of how the existing schedule is progressed. In this sense, a scheduling problem with a static nature can be included in the framework of deterministic scheduling.

On the contrary, the term "dynamic" has several construes in the literature in addition to the above definition. In a classical sense, one can say that dynamic scheduling refers to the myopic decision-making process to allocate jobs into any one of machines one after another in the manufacturing system. This means that scheduling information under dynamic context is incomplete and sometimes inaccurate unlike a situation in static scheduling. For example, the scheduler knows neither when jobs comes in the system accurately nor how long the processing time of a certain job is in advance. One may also say dynamic scheduling simply refers to the process of scheduling decisions utilizing information associated with elapsed time which may change momentarily, e.g. remaining processing times of jobs and the remaining time to the due date and so on.

These interpretations for dynamic scheduling have lead the way to develop dispatching rules or methods to update the existing schedule in realtime. Considering research directions over two decades, dynamic scheduling can also be referred to as a decision-making process to cope with uncertainty under dynamic manufacturing environments.

Offline versus Online

The interpretation of scheduling models and problems described in the above have been from academia. A different classification of scheduling is given by

(C) Offline versus Online

where the basic concept and definition of online scheduling was discussed in Sects. 1.2 and 1.3. It can be considered as a more practical concept in recent years than classifications (A) and (B). Classification (C) is also based on when to schedule.

According to the definition in Sects. 1.2 and 1.3, a decision-making process to construct shop floor schedules based on the results of short-term capacity planning is an offline scheduling activity. Offline scheduling can be mentioned basically within the framework of static scheduling, that is, the target scheduling problem is supposed to be static. Offline scheduling is, however, no more classical since it includes modern approaches of scheduling called *proactive scheduling strategies* such as *robust scheduling* [82] and *predictable scheduling* [98], where the approach of proactive scheduling is introduced to predict disruptions and unexpected events which may occur at the manufacturing phase. The proactive approaches are realistic beyond traditional deterministic approaches since the latter approaches ignore the progress of manufacturing.

Online scheduling can be roughly classified into two categories:

- *Dispatching*,
- *Schedule revision*.

In the above, dispatching is to be realized by the following two steps; (1) select a single job among a set of jobs in progress within the buffer on the basis of some suitable criteria and (2) assign it to a machine available. The criteria or rules in this approach are called *priority rules* or *dispatching rules*. The disadvantage of this approach is that decision makings will be myopic.

On the other hand, schedule revision is further classified into reactive scheduling and periodic schedule revision, both of which aim to revise the existing schedule promptly and dynamically in reaction to disruption of smooth progress of manufacturing and/or unexpected events causing significant changes in manufacturing status. The reactive scheduling is conducted on the event-driven basis, while the periodic schedule revision is carried out on the scheduled times periodically.

2.3 Generic Models for Scheduling

A great amount of research efforts have addressed deterministic models for scheduling. Over the last 50 years, in fact, a considerable number and variety of deterministic models have been developed, and then they have provided categorical specifications for a series of scheduling problem. Under the specifications, deterministic models assume that the number, n of jobs and the number, m of resources are finite and never change. Resources basically refer to workforces as well as hardwares such as machine tools. The term "machine" is often used to represent such functional resource. Actually in many theoretical models for scheduling problems, a resource to be considered is often limited to a machine. This is because a deterministic scheduling problem is traditionally referred to as a *machine scheduling problem.*

Each individual job is assumed to consists of several tasks. The following attributes of job j ($j = 1, 2, \ldots, n$) are possibly specified.

- Release date (r_j)
  The release date r_j of job j is the time where job j arrives at the manufacturing system, which is equivalent to the earliest possible time when we can start processing job j.
- Due date (d_j)
  The due date d_j of job j indicates the date when processing the job has to be completed, or equivalently the shipping date promised to the its customer.
- Processing time (p_{jk})
  The processing time p_{jk} represents the time required to process the task of job j on machine k. In stead of p_{jk}, p_j is used when job j is to be processed on a pre-specified machine or its processing time is independent of the machine.
- Weight or importance factor (w_j)
  The weight w_j expresses the relative cost or degree of importance of job j to the other jobs.

Basically, a scheduling problem is supposed to belong to a certain problem class described by a three-tuplet:

$$\alpha|\beta|\gamma$$

which is a well-known notation in deterministic scheduling [19, 23, 83, 116]. This notation can actually cover the structure of a great number of deterministic scheduling models in the literature. The first field α in this notation is specified by the machine

Fig. 2.1 An example of
the single machine models
($\alpha = 1$)

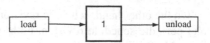

environment, i.e. the target manufacturing system, its sub system or its stage to be focused on. In field α it is allowed to specify the limited number of stages or machines if needed. The second field β describes the restrictions or constraints of the scheduling problem and also describes characteristics concerned with job processing. Finally, the third field γ expresses the criterion or criteria to be optimized.

According to the above notations, we summarize significant manufacturing configurations, restrictions and constraints, and the performance measure in the following sections.

2.4 Manufacturing Configurations

We here classify the manufacturing configurations into three models—a basic model, a typical shop model and an advanced model and then introduce various manufacturing configurations.

Basic Models

The basic models of manufacturing configurations consider a single machine and a parallel machine environment. Those are construed as the component (e.g. a specific area of the factory or a specific stage) of the whole manufacturing system.

Single machine: The single machine model as shown in Fig. 2.1 is actually the simplest machine configuration among other configurations. One may claim that the single machine model oversimplifies a scheduling problem and therefore the model tends to lack in practicality. However, it appears that the importance of the single machine model has increased in practical situations because of the progress of Flexible Manufacturing Cell (FMC) system, or a so called *cell production system* under varied-mix varied-volume production. The FMC system consists of a high performance CNC machine tool, a store and a material handling system such as a loading/unloading robot or an automated palette changer. A multifunctional machining center is also referred to as FMC.

Especially in Japan, a cell production system is recognized as one of the principles of evolved manufacturing from a multi-skilled workers together with KAIZEN activity, which has been gradually percolating through manufacturers of motor vehicles and machine tools. The cell production system is mainly adopted to an assembly process of semiconductor products [130, 139, 140], where one or a few workers

Fig. 2.2 An example of
parallel machine models

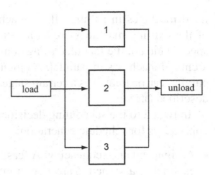

handle the all process from components mounting to inspection. The cell production
system is called a complete cell production system if the system is formed by only
one multi-skilled workers. In the so called *semi compelete* cell production system,
more than one multi-skilled workers are allocated. Single machine scheduling is
necessary for the complete cell production system. Moreover, the single machine
model has a key role of sophisticated decomposition algorithms like the Lagrangian
relaxation method [94, 150] and the shifting bottleneck procedure [66, 107, 141].

Parallel machine: The parallel machine model focuses on a specific stage which
contains m machines in parallel. Figure 2.2 illustrates the configuration of three
machines in parallel, where each job j has a single task to be processed on any
one of the m machines or on any one of m_j machines where $1 \leq m_j \leq m$. A set of
m_j machines is a specific sub set of m machines in the system on which job j can be
processed. The parallel machine environments can be classified into the following
three types according to their processing speeds [19, 116].

- Identical machines in parallel: The identical machines in parallel refers to the
  parallel machine environment where m machines are entirely the same in performance.
- Uniform machines in parallel: The uniform machines in parallel is a simple extension of the identical machines in parallel. It refers to the machine configuration,
  where m machines do not necessarily have identical processing ability [14, 71,
  77]. The processing time p_{jk} of job j on machine k is given by p_j/v_k, where v_k
  denotes the processing speed of machine k.
- Unrelated machines in parallel: The unrelated machines in parallel is a generalized
  model for the above two models [47, 90, 112]. The processing time p_{jk} of job j
  on machine k is given by p_j/v_{jk}, where each job j can be processed on machine
  k at processing speed v_{jk}. In other words the speed of the machine to process a
  certain job depends on the job.

In the real manufacturing environments, a certain area with FMCs on a shop
floor can occasionally be modeled as the machines in parallel. When the old and
the new machining centers are allocate in the same area, then the target manufacturing configuration can be modeled as uniform machines in parallel or as unre-

lated machines in parallel. If all machining centers are entirely identical (e.g. all of the existing machines are simultaneously replaced by new ones with the same specifications), the manufacturing configuration of the target area can be treated as identical machines in parallel. As mentioned above, a set of machines arranged in parallel can be treated as a certain stage in a flexible shop system which are, in detail, described later.

In regard to the scheduling decisions in the machines in parallel, there exist two approaches for schedule generation:

- Assignment and sequencing: A feasible schedule is constructed in such a way that each job is assigned to any one of machines on some suitable criterion yielding a sub set of the given jobs for each machine, and then, on each machine, the jobs are sequenced according to some sequencing algorithm. This procedure is necessarily repeated for several times until a preferable schedule is obtained.
- List scheduling: List scheduling is one of the frameworks for scheduling algorithms to determine which job should be processed next. In a deterministic sense, the list scheduling method can be described as follows:

Step 1 A so called *job list* is generated by ordering the jobs according to some rule. The job list is a permutation of all jobs. The initial completion times of all the machines are set to zero.

Step 2 The front job of the job list is allocated to any machine at time zero ($t = 0$). It is eliminated from the job list and the completion time of the selected machine is updated by adding the processing time of the assigned job.

Step 3 Likewisely, the first job of the job list is allocated to any machines having the minimum completion time. Then both the job list and the completion time of the target machine are updated.

Step 4 The above **Step 3** is repeated until all the jobs are allocated to some machine, i.e., the job list becomes empty. Note that the list scheduling algorithm can be regarded as a mapping of a job list to a feasible schedule.

Typical Shop Models

The typical shop models contain a job shop, an open shop and flow shop that are recognized as traditional models on the theoretical side [34].

Job shop: In the job shop configuration, each individual job has its own sequence of machines, by which each job is to be processed. The machine sequence of an arbitrary job is not necessarily identical to those of other jobs. The machine sequence is occasionally referred to as job routing [22, 36, 144]. Figure 2.3 depicts a job shop configuration with three machines. Over these five decades, a so called *generalized job shop scheduling problems* have been dealt with mainly within a framework of deterministic scheduling problems [23, 59] and particularly for the use for benchmark

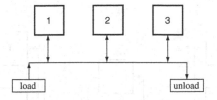

Fig. 2.3 An example of the job shop/open shop models

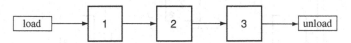

Fig. 2.4 An example of the flow shop models

tests. A generalized job shop problem refers to a job shop problem where each job is to be processed on each machine exactly once.

The practical job shop configurations include a typical FMS aiming to react flexibly to rapid changes in manufacturing environments [44, 75]. The FMS is a sophisticated Computer Aided Manufacturing (CAM) system consisting of several CNC machine tools [5] along with automated material handling devices such as AGVs or industrial robots. The FMS is also recognized as one of important hardware to realize flexible automation and computer integrated manufacturing (CIM). In FMS, there are two types of flexibility—job routing flexibility and machine flexibility. Job routing flexibility indicates the system has the ability not only to utilize multiple machines (machines in parallel) but to accept frequent changes in capability or capacity. Machine flexibility is the ability to accept changes or arrangements of orders or due dates. In contrast to this high functionality, FMS often renders production control complicated as well as scheduling decisions [16].

Open shop: In the open shop configuration, no jobs have predetermined machine sequence or route to follow. That is, there exist no precedence relations between tasks in each job [48, 86–88]. The job routing is *open*, thus the route of each job is supposed to be decided by a scheduler. The open shop problem is how we can find both a machine sequence of each job and a job order for each machine (Fig. 2.3).

Flow shop: In the flow shop environment with m machines in series, the jobs are to be processed in the same machine sequence, e.g. they have to visit machine 1, then go through machine 2, and so forth. Figure 2.4 depicts a typical flow shop configuration with three machines. In general, the model for flow shop scheduling problems is referred to as the *permutation flow shop* model on the condition that the sequence of jobs on each machine is exactly the same. There is the other type of the flow shop problems such as a *modified flow shop problem* [4, 97] and a *hybrid flow shop problem* [142] Under the modified flow shop problem, the jobs can enter one of several machines and go through a predetermined number of machines, and then exit the system from one of several machines. The hybrid flow shop consists of several manufacturing cells each of which is structured as a multi-station flow shop.

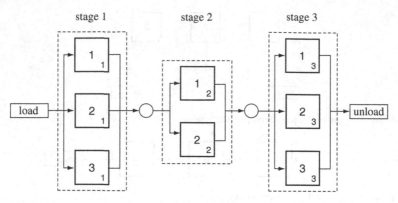

Fig. 2.5 An example of the flexible flow shop models

Advanced Shop Models

There exist some other variations in practice such as flexible flow shops [21, 35, 148] or flexible job shops [39, 122]. They are occasionally referred to as *generalized job shops* [73] or as the well-known flexible shops. The large scale FMSs have also been modeled as the flexible shop.

Flexible flow shop: A flexible flow shop is configured with c $(= 1, 2, \ldots)$ stages arranged in series as depicted in Fig. 2.5, where each stage consists of a single machine or several ones in parallel. All the jobs have an identical sequence with respect to the stages, showing that a flexible flow shop is a generalization of the flow shop. In addition, let stage θ $(= 1, 2, \ldots, c)$ has m_θ machines in parallel, and at stage θ, job j is to be processed by any machine in the stage or any one of $m_{\theta,j}$ machines in the stage where $1 \leq m_{\theta,j} \leq m_\theta$.

Flexible job shop: A flexible job shop contains c $(= 1, 2, \ldots)$ stages, each of which is of m_θ $(\theta = 1, 2, \ldots, c)$ machines in parallel as shown in Fig. 2.6. Each individual job has its own stage sequence different from those of other jobs, and this is why it is called a flexible job shop. Obviously, the flexible job shop is a the generalization of the job shop. At stage θ, each job j is to be processed by any machine in the stage or any one of $m_{\theta,j}$ machines here $1 \leq m_{\theta,j} \leq m_\theta$.

The flexible job shop is more complicated than the flexible flow shop especially in regard to the job routings and to their control.

Intriguing characteristics of the flexible shop model have been summarized in relation to job routing flexibility [22, 67, 81]. Particularly, flexibility realized by parallel machines at each stage can provide better schedule in job routings with remarkable load reduction of a bottleneck machine.

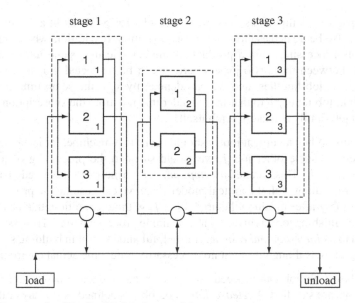

Fig. 2.6 An example of the flexible job shop models

2.5 Characteristics and Restrictions

The entries into field β are basically associated with job processing, and they play a role of constrains in many cases, making problems more specific or more complicated. In another respect, however, they make the problems more realistic. Here we introduce a significant characteristics in scheduling models related to uncertainty in scheduling decisions.

Sequence dependent setup time: When a machine finishes processing of job j ($1 \leq j \leq n$), a setup activity is required to switch to the next job denoted by $j'(\neq j)$. The time for such an activity is called *setup time*. If the setup time is dependent on the sequence of job j and j', it is called a *sequence-dependent setup time* and denoted by $s_{jj'}$. It is postulated, in many cases, that we have $s_{jj'} \neq s_{j'j}$, which makes the problem more complicated. There are numerous models for deterministic scheduling problems assuming sequence-dependent setup times [7, 105, 106, 120].

If the setup of a machine is independent of the job order, the setup time of the machine for job j can be given simply by s_j. In such a case, however, we can incorporate the setup time s_j into the processing time of job j. On the contrary, if the setup time should be specified explicitly between tasks, $s_{jj'k}$ is used to express the sequence-dependent setup time inflicted when job j is followed by job j' on machine k.

Job Families: The setup time can concisely be expressed by introducing a concept of job families [124]. It is inseparably related to the product families, where products in the same family are basically homogenous in structural characteristics to share common parts or components, and therefore, jobs for products in a same product family can also be treated as those in a same job family. In addition, when a machine switches its processing from a product to another one in a same job family, the setup time between these two jobs is zero or can be neglected. Suppose we have $Y(= 1, 2, \ldots)$ job families, and we can denote, by $s_{yy'}$, the setup time to switch form a job in job family y to that in y'. Under this notation, the subscription y or y' signifies a job family instead of a job itself.

Preemptions: When an urgent job j_u should occur to a machine, for instance, while it is processing some other job j, we would suspend the processing of job j to switch from j to j_u. This is called *preemption*. Furthermore, it is assumed, in most of the literature dealing with theoretical models, that we can resume the processing of the former job j after finishing the urgent job j_u without condition, that is, the time required for finishing job j is given by its remaining processing time. This assumption is referred to as *interrupt and resume*. It is helpful and skillful in building stochastic scheduling models though it might not necessarily reflect the actual environments.

Release dates: No job can proceed on to the processing stage before the relevant requirements are satisfied. A *release date* r_j of job j is defined by the day or the time after which j can be transferred in its manufacturing process. Unless r_j is exactly known in advance, the problem is, in many cases, discussed within a framework of stochastic scheduling problems.

Breakdowns: If a machine breakdown occurs, the processing of any job on the breakdown machine is naturally aborted. The job cannot be processed until the machine is recovered from its breakdown. The time between a machine breakdown and its recovery is called *downtime*. The downtime is substantially equivalent to the prohibited period during which any job cannot be loaded into the machine. In this sense, the maintenance for a machine can be dealt with as the machine breakdown.

2.6 Performance Measure

The objective function to be optimized in a scheduling problem is inseparable from completion times of the jobs. Let C_{jk} express the completion time of a task of job j on machine k, and C_j signify the completion time job j in the manufacturing system. If needed, we can also, by notating $C_{jk}(S)$ or $C_j(S)$, specify schedule S.

Performance measures focusing on completion times of the jobs are as follows:

Total Weighted Completion Times: $\sum w_j C_j$

One of useful measures is the aggregate amount of completion times of jobs. Its general form is given by

$$C_{\text{sum}} = \sum_{j=1}^{n} w_j C_j, \tag{2.1}$$

where $w_j (\geq 0)$ is a weight for the completion time of job j. This measure is called total weighted completion time [72, 74, 76] in the following.

Makespan (The Length of the Schedule): $C_{\max}$

The makespan is also useful and used commonly in the literature [78, 135, 138]. It is given by

$$C_{\max} = \max(C_1, C_2, \ldots, C_n). \tag{2.2}$$

This performance measure is important since the sooner we can accomplish a schedule, the higher performance we can expect from the system. In the literature, this measure is frequently introduced for the purpose of evaluating performances of the proposed scheduling algorithms. A notation $C_{\max}(S)$ is used to represent the makespan of a schedule S. It is, however, indicated empirically that costs for work-in-process inventories might incease in reducing the makespan.

In the performance evaluation of a newly generated schedule, due dates of individual jobs are occasionally incorporated in the measure. Representative indexes associated with the due dates are *tardiness* and *earliness* of jobs. The tardiness of job j is given by

$$T_j = \max(C_j - d_j, 0). \tag{2.3}$$

It is trivial in the above that T_j is non-decreasing in C_j. The tardiness of job j will deeply influence upon customer satisfaction with j and therefore its relevant penalty cost.

The earliness of job j is defined by

$$E_j = \max(d_j - C_j, 0). \tag{2.4}$$

In the above, E_j is non-increasing in C_j of job j. In the practical situations, the earliness of jobs will be useful for an assessment of manufacturing efficiency, e.g., in extracting production surplus.

In addition to the above indexes, *lateness* of a job is possibly considered. The lateness of job j is defined by

$$L_j = C_j - d_j. \tag{2.5}$$

Obviously, the tardiness takes a negative value if the corresponding job is processed before its due date.

We further introduce an indicator to express whether or not a job is tardy, which is referred to as an *unit penalty*. The unit penalty of job j is given by

$$U_j = \begin{cases} 1 & \text{if } C_j > d_j \\ 0 & \text{otherwise} \end{cases}. \tag{2.6}$$

Typical performance measures associated with the due dates are:

Total Weighted Tardiness: $\sum w_j T_j$

The aggregate amount of tardiness of jobs is one of performance measures, which are closely related to due dates of jobs. Its general form is

$$T_{\text{sum}} = \sum_{j=1}^{n} w_j T_j, \tag{2.7}$$

where $w_j (\geq 0)$ expresses a weight for the tardiness of job j. This measure is called total weighted tardiness [11, 68, 100] in the following. It is practically important since it has an association with customer satisfaction.

Maximum Tardiness: T_{max}

The maximum tardiness [52, 95, 126] is used for grasping the worst violation of the commitment. It is given by

$$T_{\text{max}} = \max(T_1, T_2, \ldots, T_n). \tag{2.8}$$

Number of Tardy Jobs: U_{sum}

The number of tardy jobs is relatively simple and easily calculated. It is given by

$$U_{\text{sum}} = \sum_{j=1}^{n} U_j. \tag{2.9}$$

This performance measure will be stricter in practice since the degree of tardiness is less important than its presence [30, 33, 61].

Now let us allow to illustrate intriguing properties of the above three measures T_{max}, T_{sum} and U_{sum} by presenting simple examples where $w_j = 1$ for all j. Consider that, for a given scheduling problem with four jobs, two schedules denoted by S_1

and S_2 were generated, and either S_1 or S_2 is to be released into the system. Suppose that the tardiness of each job on the two schedules S_1 and S_2 is given by

$$S_1 : (T_1, T_2, T_3, T_4) = (1, 1, 1, 1),$$
$$S_2 : (T_1, T_2, T_3, T_4) = (0, 0, 4, 0),$$

then which schedule should be chosen if we introduce above performance measures? The performance measures we have observed provide:

- The maximum tardiness;

$$T_{max}(S_1) = 1, \ T_{max}(S_2) = 4$$

- The total tardiness;
$$T_{sum}(S_1) = 4, \ T_{sum}(S_2) = 4$$

- The number of tardy jobs;

$$U_{sum}(S_1) = 4, \ U_{sum}(S_2) = 1$$

Obviously the total tardiness cannot point out the difference between schedules S_1 and S_2 because their total tardiness are identical. If one should compare the two schedules through their maximum tardiness, schedule S_1 would be preferable. Under schedule S_1, however, none of the jobs meet the due dates resulting in customers' dissatisfaction. This indicates that the number U_{sum} of tardy jobs could be one of major performance measures.

Total Weighted Earliness and Tardiness: ET

The aggregate amount of both earliness and tardiness of jobs would also be a performance measure. Its general form is given by

$$ET = \sum_{j=1}^{n} \left(w_j^{(1)} E_j + w_j^{(2)} T_j \right), \tag{2.10}$$

where $w_j^{(1)}$ and $w_j^{(2)}$ signify the weight associated with the earliness of job j and that associated with the tardiness of job j, respectively. In scheduling theory, the idea of Just In Time (JIT) can be modeled as a deterministic scheduling problem minimizing weighted job earliness and tardiness [69, 118]. One of the reason why this "JIT scheduling problem" is subjected to study for many years is that JIT production became popular along with improvements in infrastructure involving production information systems. In the JIT scheduling problems, a scheduler attempts to an insert idle time intentionally to the schedule in order to minimize the deviation between the completion time and the due date of the job as much as possible, thus the optimal starting time of a job is not always equivalent to the earliest time of that job. A similar measure to the above has been discussed by Kolahan [69] who introduced total setups in order to keep continuity of job processing together with earliness and tardiness penalties. He also discussed the trade-off between continuity and timeliness of job processing.

If the performance measure is non-decreasing in completion times $C_1, C_2, \ldots, C_n$ of all jobs, then it is called *regular*, and otherwise it is called *nonregular*. The makespan $C_{\max}$, the total weighted completion time $\sum w_j C_j$, the maximum lateness $L_{\max}$, the number of tardy jobs U_j are typical examples of regular measures, while the total weighted earliness and tardiness, ET, belongs to the class of non-regular performance measure.

Other noteworthy performance measure would be the total setup times. Total setup time, denoted by TS, is given by

$$TS = \sum u_{jj'k} s_{jj'k}, \tag{2.11}$$

where $u_{jj'k}$ is a indicator variable defined by

$$u_{jj'k} = \begin{cases} 1, & \text{if the task of job } j \text{ on machine } k \text{ is immediately} \\ & \text{followed by the task of job } j' \text{ on the same machine.} \\ 0, & \text{otherwise.} \end{cases} \tag{2.12}$$

The minimum total setup time will be usually an indication for a good makespan [116].

In practice, a scheduler often considers two or more performance measures. This type of decision-making is generically called *multiobjective scheduling* or *multicriteria scheduling* [15, 137].

2.7 Online Scheduling Problems

Biographical survey on online scheduling reveals that most of literature define "offline" by the condition that all the jobs information such as the total number of jobs, their release dates and processing times, is fully known in advance. On the other hand, they define "online" by the conditions that jobs arrive one by one to be assigned to machines irrevocably as soon as they come, and that, prior to job arrivals, no information is available on the subsequent jobs (e.g., [6, 51, 114, 117]). This seems to be because online scheduling has historically and traditionally been discussed both in computer science (e.g., [26, 27, 132–134, 149]) as well as in production engineering (e.g., [55, 85, 101, 131]), simultaneously. Moreover, it is because the theoretical aspects of online scheduling have been focused on for the purpose of clarifying the characteristics of the problems and exploring methodologies to move to the optimality [8, 13, 37, 43, 136].

When we confine ourselves to production engineering, the above definition on online scheduling will restrict our interests to some extended and elaborate dispatching rules. As we have observed in Sects. 1.2 and 1.3, the terminology "online" in this book emphasizes capability of revising or adjusting the existing schedule in real time, manually or by some specific computer software, to cope with various uncertainties caused by machine breakdowns, arrivals of urgent jobs, and under- or over-estimates of setup times.

References

1. Abdelmaguid TF, Nassef AO, Kamal BA, Hassan MF (2004) A hybrid GA/heuristic approach to the simultaneous scheduling of machines and automated guided vehicles. Int J Prod Res 42(2):267–281
2. Adenso-Díaz B (1992) Restricted neighborhood in the tabu search for the flowshop problem. Eur J Oper Res 62:27–37
3. Aiex RM, Binato S, Resende MGC (2003) Parallel GRASP with path-relinking for job shop scheduling. Parallel Comput 29(4):393–430
4. Aktürk MS, Gürel S (1999) Match-up scheduling under a machine breakdown. Eur J Oper Res 112(1):81–97
5. Aktürk MS, Ilhan T (2011) Single CNC machine scheduling with controllable processing times to minimize total weighted tardiness. Comput Oper Res 38(4):771–781
6. Albers S (2010) Online scheduling. In: Robert Y, Vivien F (eds) Introduction to scheduling. Chapman and Hall/CRC, Boca Raton, pp 51–73
7. Allahverdi A, Gupta JND, Aldowaisan T (1999) A review of scheduling research involving setup considerations. Omega Int J Manag Sci 27:219–239
8. Ambühl C, Mastrolilli M (2005) On-line scheduling to minimize max flow time: an optimal preemptive algorithm. Oper Res Lett 33(6):597–602
9. Anghinolfi D, Paolucci M (2009) A new discrete particle swarm optimization approach for the single-machine total weighted tardiness scheduling problem with sequence-dependent setup times. Eur J Oper Res 193(1):73–85
10. Anwar MF, Nagi N (1998) Integrated scheduling of material handling and manufacturing activities for just-in-time production of complex assemblies. Int J Prod Res 36(3):653–681
11. Asano M, Ohta H (2002) A heuristic for job shop scheduling to minimize total weighted tardiness. Comput Ind Eng 42(2–4):137–147
12. Ausiello G, Crescenzi P, Gambosi G, Kann V, Marchetti-Spaccamela A, Protasi M (2003) Complexity and approximation: combinatorial optimazation problems and their approximability properties. Springer, Berlin
13. Avidor A, Azar Y, Sgall J (2001) Ancient and new algorithms for load balancing in the l_p norm. Algorithmica 29:422–441
14. Azizoglu M, Kirca O (1997) On the minimization of total weighted flow time with identical and uniform parallel machines. Eur J Oper Res 113:91–100
15. Bagchi TP (1999) Multiobjective scheduling by genetic algorithm. Kluwer Academic Publishers, Boston
16. Ben-Daya M (1995) FMS short term planning problems: a review. Manuf Res Technol 23:113–139
17. Ben-Daya M, Al-Fawzan M (1998) A tabu search approach for the flow shop scheduling problem. Eur J Oper Res 109:88–95

18. Bierwirth C, Mattfeld DC (1999) Production scheduling and rescheduling with genetic algorithm. Evol Comput 7(1):1–17
19. Błażewicz J, Ecker KH, Pesch E, Schmidt G, Weglarz (1996) Scheduling computer and manufacturing processes. Springer, Berlin
20. Blum C (2005) Ant colony optimization: introduction and recent trends. Phys Life Rev 2(4):353–373
21. Brah SA, Loo LL (1999) Heuristics for scheduling in a flow shop with multiple processors. Eur J Oper Res 113:113–122
22. Brandimarte P (1999) Exploiting process plan flexibility in production scheduling: a multi-objective approach. Eur J Oper Res 114:59–71
23. Brucker P (2001) Scheduling algorithm. Springer, Berlin
24. Brucker P, Knust S (2003) Complex scheduling. Springer, Berlin
25. Candido MA, Khator SK, Barcia RM (1998) A genetic algorithm based procedure for more realistic job shop scheduling problems. Int J Prod Res 36(12):3437–3457
26. Cao Q, Liu Z (2010) Online scheduling with reassignment on two uniform machines. Theor Comput Sci 411(31–33):2890–2898
27. Chandra AK, Wong CK (1975) Worst-case analysis of a placement algorithm related to storage allocation. SIAM J Comput 4(3):249–263
28. Cheng R, Gen M, Tsujimura Y (1999) A tutorial survey of job-shop scheduling problems using genetic algorithms-I representation. Comput Ind Eng 36(2):343–364
29. Croce FD, Tadei R, Volta G (1995) A genetic algorithm for the job shop problem. Comput Oper Res 22:15–24
30. Croce FD, Gupta JND, Tadei R (2000) Minimizing tardy jobs in a flowshop with common due date. Eur J Oper Res 120(2):375–381
31. Dallery Y, Bihan HL (1999) An improved decomposition method for the analysis of production lines with unreliable machines and finite buffers. Int J Prod Res 37(5):1093–1117
32. De S, Lee A (1998) Towards a knowledge-based scheduling system for semiconductor testing. Int J Prod Res 36(4):1045–1073
33. De P, Ghosh JB, Wells CE (1991) On the minimization of the weighted number of tardy jobs with random processing times and deadline. Comput Oper Res 18(5):457–463
34. Demirkol E, Mehta S, Uzsoy R (1998) Benchmarks for shop scheduling problems. Eur J Oper Res 109:141–173
35. Dessouky MM, Dessouky MI, Verma SK (1998) Flowshop scheduling with identical jobs and uniform parallel machines. Eur J Oper Res 109:620–631
36. Diallo M, Pierreval H, Quilliot A (2001) Manufacturing cells design with flexible routing capability in presence of unreliable machines. Int J Prod Econ 74:175–182
37. Du D (2009) Optimal preemptive semi-online scheduling on two uniform processors. Inf Process Lett 92(5):219–223
38. Elmaghraby SE (2010) Stochastic scheduling. Cambridge University Press, New York
39. Fattahi P, Fallahi A (2010) Dynamic scheduling in flexible job shop systems by considering simultaneously efficiency and stability. CIRP J Manuf Sci Technol 2(2):114–123
40. Feo TA (1995) Greedy randomized adaptive search procedures. J Glob Optim 6:109–134
41. Feo TA, Bard JF, Holland SD (1996) A GRASP for scheduling printed wiring board assembly. IIE Trans 28(2):155–165
42. Feo TA, Venkatraman K, Bard JF (2003) A GRASP for a difficult single machine scheduling problem. Comput Oper Res 18(8):635–643
43. Fiat A, Woeginger G (1998) Competitive analysis of algorithms. Lect Notes Comput Sci 1442:1–12
44. Gargeya VB, Deane RH (1999) Scheduling in the dynamic job shop under auxiliary resource constraints: a simulation study. Int J Prod Res 37(12):2817–2834
45. Gawiejnowicz S (2008) Time-dependent scheduling. Springer, Berlin
46. Gershwin SB (1987) An efficient decomposition method for the approximate evaluation of tandem queues with finite storage space and blocking. Oper Res 35:291–305

47. Glass CA, Potts CN, Shade P (1994) Unrelated parallel machine scheduling using local search. Math Comput Model 20(2):41–52
48. Guéret C, Prins C (1998) Classical and new heuristics for the open-shop problem: a computational evaluation. Eur J Oper Res 107:306–314
49. Guéret C, Jussein N, Prins C (2000) Using intelligent backtracking to improve branch-and-bound methods: an application to open-shop problems. Eur J Oper Res 127:344–354
50. Hastings NAJ, Yeh CH (1990) Job oriented production scheduling. Eur J Oper Res 47:35–48
51. He Y, Dòsa G (2005) Semi-online scheduling jobs with tightly-grouped processing times on three identical machines. Discrete Appl Math 150(1–3):140–159
52. Hochbaum DS, Shamir R (1989) An $O(n \log^2 n)$ algorithm for the maximum weighted tardiness problem. Inf Process Lett 4:215–219
53. Huang SC, Lin JT (1998) An interactive scheduler for a wafer probe centre in semiconductor manufacturing. Int J Prod Res 36(7):1883–1900
54. Hussain MF, Joshi SB (1999) Job-shop scheduling—using component packing and simulated annealing approach. Int J Prod Res 37(16):3711–3723
55. Imreh C (2009) Online scheduling with general machine cost functions. Discrete Appl Math 157(9):2070–2077
56. Jackson JR (1955) Scheduling a production line to minimize maximum tardiness, Research report, vol 43, Management Science Research Project, University of California, Los Angeles
57. Jackson JR (1956) An extension of Johnson's results on job lot scheduling. Naval Res Logistics Q 3:201–203
58. Jain AK, Elmaraghy HA (1997) Production scheduling/rescheduling in flexible manufacturing. Int J Prod Res 35(1):281–309
59. Jain AS, Meeran S (1999) Deterministic job-shop scheduling: past, present and future. Eur J Oper Res 113:390–434
60. Jeong KC, Kim YD (1998) A real-time scheduling mechanism for a flexible manufacturing system: using simulation and dispatching rules. Int J Prod Res 36(9):2609–2626
61. John TC, Wu Y-B (1987) Minimum number of tardy jobs in single machine scheduling with release dates—an improved algorithm. Comput Ind Eng 12(3):223–230
62. Johnson SM (1954) Optimal two and three-stage production schedules with setup times included. Naval Res Logistics Q 1:61–67
63. Kaskavelis CA, Caramanis MC (1998) Efficient Lagrangian relaxation algorithms for industry size job-shop scheduling problems. IIE Trans 30:1085–1097
64. Kennedy J, Eberhart R (1995) Particle swarm optimization. In: Proceeding of the 1995 IEEE international conference on neural network, pp 1942–1948
65. Kerr R, Szelke E (1995) Artificial intelligence in reactive scheduling. Chapman and Hall, London
66. Kim JU, Kim YD (1999) A decomposition approach to a multi-period vehicle scheduling problem. Omega Int J Manag Sci 27:421–430
67. Kim K-H, Egbelu PJ (1999) Scheduling in a production environment with multiple process plans per job. Int J Prod Res 37(12):2725–2753
68. Kim D-W, Na D-G, Chen FF (2003) Unrelated parallel machine scheduling with setup times and a total weighted tardiness objective. Robot Comput Integr Manuf 9(1–2):173–181
69. Kolahan F, Liang M (1998) An adaptive TS approach to JIT sequencing with variable processing times and sequence-dependent setups. Eur J Oper Res 109:142–159
70. Kolonko M (1999) Some new results on simulated annealing applied to the job shop scheduling problem. Eur J Oper Res 113:123–136
71. Koulamas C, Kyparisis GJ (2000) Scheduling on uniform parallel machines to minimize maximum lateness. Oper Res Lett 26(4):175–179
72. Kovács A, Beck JC (2008) A global constraint for total weighted completion time for cumulative resources. Eng Appl Artif Intell 21(5):691–697
73. Krüger K, Sotskov YN, Werner F (1998) Heuristics for generalized shop scheduling problems based on decompositions. Int J Prod Res 36(11):3013–3033

74. Kubiak W, Timkovsky V (1996) Total completion time minimization in two-machine job shops with unit-time operations. Eur J Oper Res 94(2):310–320
75. Kutanoglu E, Sabuncuoglu I (1999) An analysis of heuristics in a dynamic job shop with weighted tardiness objectives. Int J Prod Res 37(1):165–187
76. Kyparisis GJ, Koulamas C (2001) A note on weighted completion time minimization in a flexible flow shop. Oper Res Lett 29(1):5–11
77. Kyparisis GJ, Koulamass C (2006) Flexible flow shop scheduling with uniform parallel machines. Eur J Oper Res 168(3):985–997
78. Kyparisis GJ, Koulamas C (2006) A note on makespan minimization in two-stage flexible flow shops with uniform machines. Eur J Oper Res 175(2):1321–1327
79. Laarhoven PJMV, Aarrts EHL, Lenstra JK (1992) Job shop scheduling by simulated annealing. Oper Res 40(1):113–125
80. Lawler EL et al (1993) Sequencing and scheduling: algorithms and complexity. In: Graves SC et al (eds) Logistics of production and inventory, North-Holland, Amsterdam, pp 445–522
81. Lee C-Y, Vairaktarakis GL (1998) Performance comparison of some classes of flexible flow shops and job shops. Int J Flex Manuf Syst 10:379–405
82. Leon VJ, Wu SD, Storer RH (1994) Robustness measures and robust scheduling for job shops. IIE Trans 26(5):32–43
83. Leung JYT (2004) Introduction and notation. In: Leung JYT (ed) Handbook of scheduling. Chapman and Hall/CRC, Boca Raton
84. Li D-C, Lin H-K, Torng K-Y (1996) A strategy for evolution of algorithms to increase the computational effectiveness of NP-hard scheduling problems. Eur J Ope Res 88:404–412
85. Li R, Huang H-C (2007) Improved algorithm for a generalized on-line scheduling problem on identical machines. Eur J Oper Res 176(1):643–652
86. Liaw CF (1998) An iterative improvement approach for the nonpreemptive open shop scheduling problem. Eur J Oper Res 111:509–517
87. Liaw CF (1999) Applying simulated annealing to the open shop scheduling problem. IIE Trans 31:457–465
88. Liaw CF (2000) A hybrid genetic algorithm for the open shop scheduling problem. Eur J Oper Res 124:28–42
89. Lin TL, Horng SJ, Kao TW, Chen YH, Run RS, Chen RJ, Lai JL, Kuo IH (2010) An efficient job-shop scheduling algorithm based on particle swarm optimization. Expert Syst Appl 37(3):2629–2636
90. Lin YK, Pfund ME, Fowler JW (2011) Heuristics for minimizing regular performance measures in unrelated parallel machine scheduling problems. Comput Oper Res 38(6):901–916
91. Liu J, MacCarthy BL (1999) General heuristic procedures and solution strategies for FMS scheduling. Int J Prod Res 37(14):3305–3333
92. Lopez L, Carter MW, Gendreau M (1998) The hot strip mill production scheduling problem: a tabu search approach. Eur J Oper Res 106:317–335
93. Luh PB, Hoitomt DJ, Max E, Pattipati KR (1990) Schedule generation and reconfiguration for parallel machines. IEEE Trans Robot Autom 6(6):687–696
94. Luh PB, Hoitomt DJ (1993) Scheduling of manufacturing systems using Lagrangian relaxation technique. IEEE Trans Autom Control 38(7):1066–1088
95. Luo X, Chu C (2007) A branch-and-bound algorithm of the single machine schedule with sequence-dependent setup times for minimizing maximum tardiness. Eur J Oper Res 180(1):68–81
96. Mattfeld DC (1996) Evolutionary search and the job shop. Physica, Heidelberg
97. McPherson RF, White KP Jr (1998) Periodic flow line scheduling. Int J Prod Res 36(1):51–73
98. Mehta SV, Uzsoy RM (1998) Predictable scheduling of a job shop subject to breakdowns. IEEE Trans Robot Autom 14(3):365–378
99. Morton TE, Pentico DW (1993) Heuristic scheduling systems. Wiley, New York
100. Naderi B, Zandieh M, Shirazi MAHA (2009) Modeling and scheduling a case of flexible flowshops: total weighted tardiness minimization. Comput Ind Eng 57(4):1258–1267

101. Nong Q, Yuan J, Fu R, Lin L, Tian J (2008) The single-machine parallel-batching on-line scheduling problem with family jobs to minimize makespan. Int J Prod Econ 111(2):435–440
102. Noronha SJ, Sarma VVS (1991) Knowledge-based approaches for scheduling problems: a survey. IEEE Trans Knowl Data Eng 3(2):160–171
103. Nowicki E (1999) The permutation flow shop with buffers: a tabu search approach. Eur J Oper Res 116:205–219
104. Oliff MD (1988) Expert systems and intelligent manufacturing. North-Holland, New York
105. Ovacik IM, Uzsoy R (1994) Rolling horizon algorithms for a single-machine dynamic scheduling problem with sequence-dependent setup times. Int J Prod Res 32(6):1243–1263
106. Ovacik IM, Uzsoy R (1995) Rolling horizon procedures for dynamic parallel machine scheduling with sequence-dependent setup times. Int J Prod Res 33(11):3173–3192
107. Ovacik IM, Uzsoy R (1997) Decomposition methods for complex factory scheduling problems. Kluwer Academic Publishers, Massachusetts
108. Park M, Kim Y (2000) A branch and bound algorithm for a production scheduling problem in an assembly system under due date constraints. Eur J Oper Res 123:504–518
109. Perregaard M, Clausen J, Parallel branch-and-bound methods for the job-shop scheduling problem. Annals Oper Res 83:137–160
110. Pezzella F, Merelli E (2000) A tabu search method guided by shifting bottleneck for the job shop scheduling problem. Eur J Oper Res 120:297–310
111. Pflughoeft KA, Hutchinson GK, Nazareth DL (1996) Intelligent decision support for flexible manufacturing: design and implementation of a knowledge-based simulator. Omega Int J Manag Sci 24(3):347–360
112. Piersma N, Dijk WV (1996) A local search heuristic for unrelated parallel machine scheduling with efficient neighborhood search. Math Comput Model 24(9):11–19
113. Pinedo M, Weiss G (1984) Scheduling jobs with exponentially distributed processing times o two machine with resource constraints. Manag Sci 30(7):883–889
114. Pinedo M (2004) Offline deterministic scheduling, stochastic scheduling, and online deterministic scheduling: a comparative review. In: Leung JY-T (ed) Handbook of scheduling scheduling—algorithms, models, and performance analysis. Chapman and Hall/CRC, Boca Raton, pp 38-1– 38-14
115. Pinedo M (2005) Planning and scheduling in manufacturing and services. Springer Science+Business Media, New York
116. Pinedo M (2008) Scheduling—theory, algorithms, and systems. Springer, New York
117. Pruhs K, Sgall, J Torng E (2004) Online scheduling. In: Leung JY-T (ed) Handbook of scheduling scheduling—algorithms, models, and performance analysis. Chapman and Hall/CRC, Boca Raton, pp 15-1–15-39
118. Rajendran C (1999) Formulations and heuristics for scheduling in a Kanban flowshop to minimize the sum of weighted flowtime, weighted tardiness and weighted earliness of containers. Int J Prod Res 37(5):1137–1158
119. Resende MGC, Sousa JP (2004) Metaheuristics: computer decision making. Kluwer Academic Publishers, London
120. Ríos-Mercado RZ, Bard JF (1999) A branch-and-bound algorithm for permutation flow shops with sequence-dependent setup times. IIE Trans 31:721–731
121. Rossi A, Dini G (2000) Dynamic scheduling of FMS using a real-time genetic algorithm. Int J Prod Res 38(1):1–20
122. Rossi A, Dini G (2007) Flexible job-shop scheduling with routing flexibility and separable setup times using ant colony optimisation method. Robot Comput Integr Manuf 23(5):503–516
123. Sabuncuoglu I (1998) A study of scheduling rules of flexible manufacturing systems: a simulation approach. Int J Prod Res 36(2):527–546
124. Schaller J (2000) A comparison of heuristics for family and job scheduling in a flow-line manufacturing cell. Int J Prod Res 38(2):287–308
125. Shanker K, Modi BK (1999) A branch and bound based heuristic for multi-product resource constrained scheduling problem in FMS environment. Eur J Oper Res 113:80–90

126. Shanthikumar JG (1983) Scheduling n jobs on one machine to minimize the maximum tardiness with minimum number. Comput Oper Res 10(3):255–266
127. Smith JS, Peters BA, Srinivasan A (1999) Job shop scheduling considering material handling. Int J Prod Res 37(7):1541–1560
128. Steiner G, Yeomans JS (1994) Optimal level schedules in mixed-model, multi-level JIT assembly systems with pegging. Eur J Oper Res 95(1):38–52
129. Steinhőfel K (1999) Stochastic algorithm in scheduling theory. Infix, Sankt Augustin
130. Sung CS, Choung YI (2000) Minimizing makespan on a single burn-in oven in semiconductor manufacturing. Eur J Oper Res 120:559–574
131. Tan Z, He Y (2001) Semi-on-line scheduling with ordinal data on two uniform machines. Oper Res Lett 28(5):221–231
132. Tan Z, He Y, Epstein L (2005) Optimal on-line algorithms for the uniform machine scheduling problem with ordinal data. Inf Comput 196(1):57–70
133. Tan Z, He Y (2007) Semi-online scheduling problems on two identical machines with inexact partial information. Theor Comput Sci 377(1–3):110–125
134. Tao J, Chao Z, Xi Y, Tao Y (2010) An optimal semi-online algorithm for a single machine scheduling problem with bounded processing time. Inf Process Lett 110(8–9):325–330
135. Tavakkoli-Moghaddam R, Daneshmand-Mehr M (2005) A computer simulation model for job shop scheduling problems minimizing makespan. Comput Ind Eng 48(4):811–823
136. Tian J, Fu R, Yuan J (2009) A best online algorithm for scheduling on two parallel batch machines. Theor Comput Sci 410(21–23):2291–2294
137. T'kindt V, Billaut JC (2002) Multicriteria scheduling—theory, models and algorithms. Springer, Berlin
138. Toksari MD (2011) A branch and bound algorithm for minimizing makespan on a single machine with unequal release times under learning effect and deteriorating jobs. Comput Oper Res 38(9):1361–1365
139. Uzsoy R, Lee CY, Martin-Vega LA (1992) A review of production planning and scheduling models in the semiconductor industry part I: system characteristics, performance evaluation and production planning. IIE Trans 24:47–60
140. Uzsoy R, Lee CY, Martin-Vega LA (1994) A review of production planning and scheduling models in the semiconductor industry Part II: shop-floor control. IIE Trans 26:44–55
141. Uzsoy R, Wang CS (2000) Performance of decomposition procedures for job shop scheduling problems with bottleneck machines. Int J Prod Res 38(6):1271–1286
142. Vairaktarakis G, Elhafsi M (2000) The use of flowlines to simplify routing complexity in two-stage flowshops. IIE Trans 32:687–699
143. Valls V, Perez MA, Quintanilla MS (1998) A tabu search approach to machine scheduling. Eur J Oper Res 106:277–300
144. Wahab MIM, Stoyan SJ (2008) A dynamic approach to measure machine and routing flexibilities of manufacturing systems. Int J Prod Econ 113(2):895–913
145. Watanabe T, Sakamoto M (1985) On-line scheduling for adaptive control machine tools in FMS. Robot Comput Integr Manuf 2(1):65–73
146. Wiers VCS (1997) A review of the applicability of OR and AI scheduling techniques in practice. Omega Int J Manag Sci 25(2):145–153
147. Yagmahan B, Yenissey MM (2008) Ant colony optimization for multi-objective flow shop scheduling problem. Comput Ind Eng 54(3):411–420
148. Yang Y, Kreipl S, Pinedo M (2000) Heuristics for minimizing total weighted tardiness in flexible flow shops. J Sched 3:89–108
149. Zhang G (1997) A simple semi on-line algorithm for $P_2||C_{max}$ with a buffer. Inf Process Lett 61(14):145–148
150. Zhang Y, Luh PB, Yoneda K, Kano T, Kyoya Y (2000) Mixed-model assembly line scheduling using the Lagrangian relaxation technique. IIE Trans 32:125–134
151. Zhang G, Shao X, Li P, Gao L (2009) An effective hybrid particle swarm optimization algorithm for multi-objective flexible job-shop scheduling problem. Comput Ind Eng 56(4): 1309–1318

Part II
Online Scheduling Models

Chapter 3
When-to-Revise Against Uncertainty?

Abstract Even if a pertinent schedule is initially obtained, its smooth progress will be inhibited by a variety of uncertainties in the real circumstances. It is, as a matter of course, very important to provide against these uncertainties, but at the same time the proactive activities against them have limitations after all. One of the realizable approaches to this kind of problem is to adaptively revise the ongoing schedule or make an adjustment to it on a case-by-case basis. This chapter focuses on the timing to revise or adjust the existing schedule on the condition that the status of the ongoing schedule is monitored. Considered are four types of policies all of which prescribe the timing of revising the ongoing schedule.

3.1 Uncertainty in Scheduling

3.1.1 Typology of Uncertainty

From the academic standpoint, Aytug et al. [3] have attempted to classify uncertainties observed in manufacturing in terms of the following four dimensions:

(1a) *Cause*;
(1b) *Context*;
(1c) *Impact*;
(1d) *Inclusion*.

They view the dimension *cause* as *object* and *state*. The object is represented by resource, worker, material, process, facility and tooling, while the state is by a qualitative nature e.g., in-process, idle, ready, not ready, damaged. They also pointed out that researchers addressing scheduling with uncertainty focus on resource oriented uncertainty, e.g., varying processing times of jobs, mean time between failures, and mean time to repair, all of which can explicitly be modeled based on probability distributions.

H. Suwa and H. Sandoh, *Online Scheduling in Manufacturing*,
DOI: 10.1007/978-1-4471-4561-5_3, © Springer-Verlag London 2013

The dimension *context* is synonymous with the environmental situation at the time when individual planned events occur. The situation is either *context-free* or *context-sensitive*. A *context-free* situation would require no additional information or specific decision makings, while a *context-sensitive* situation would have information about the context and its implication.

Impact refers to the outcome of uncertainty that is categorized as time, material, quality, independent or dependent, and context-free or context-sensitive, where *independent* and *dependent* refer to possible relationships to other jobs and internal/external activities. For example, deviations of the starting and completion time for job/task processing or setup would be classified into *time*. In case the uncertainty impacted the availability of a certain material, it would be categorized as *material*.

The above three dimensions are considered to be on the problem side, while the fourth dimension *inclusion* is on the methodological or problem-solving side. This dimension is relate to the existence of some suitable dynamics to cope with uncertainties. In the scheduling environments, there are two ways to accommodate uncertainties: one is to cope with uncertainties in advance during the process of offline scheduling, which would be construed as a priori action. The other is to react as promptly as possible after contingencies due to uncertainties, which would be referred to as a posteriori action. The framework of proactive scheduling is a typical example of the former since it includes robust scheduling or the slack protection/match-up strategy, while the reactive scheduling methods are examples of the latter.

McKay et al. [22] have shown another perspective by introducing the following three categories for uncertainties:

(2a) *Complete unknowns*;
(2b) *Suspicions* about the future of the decision-maker;
(2c) *Known uncertainties*.

Complete unknowns refer to unanticipated events which cannot be dealt with on the shop floor and caused by external factors. Information about them are not available in advance, e.g., Toyota underwent supply disruptions owing to the fire of a factory (1997), strikes by labor unions in China (2010) and the huge earthquake (2011).

Suspicions about the future arise from the so-called *implicit knowledge* of experts. They are generally recognized as empirical knowledge and intuition which are considered indispensable for human's decisions. For this reason, a wide variety of knowledge base systems where such knowledge was incorporated were once developed, but as pointed out in Sect. 2.1, it has been left unfulfilled since it is very difficult to describe the knowledge explicitly and quantitatively.

The last type of uncertainties, known uncertainties, are the ones about which some information is obtainable during the scheduling process. Behaviors of unexpected events with known uncertainties can be described by means of some specific stochastic models, e.g., the number of unexpected events over per unit of time can be described by employing a suitable probability distribution. In other words, *known uncertainties* are the only type of uncertainties for which we have some methods in the process of scheduling.

3.1.2 Known Uncertainties

This subsection confines itself to known uncertainties to see them in great detail. The known uncertainties are realized in one of the following three types:

(3a) *Disruption*;
(3b) *Interruption*;
(3c) *Variation*.

Disruption refers to the uncertainties mainly related to machines, and events in this category are tool breakages [27], tool change delays [13], sudden machine breakdowns [1, 2, 21], workers' accidents [31], and so forth (e.g., [15, 20]). A disruption incurs a certain period on the existing schedule within which job processing is not allowed or prohibited, and therefore we are forced to change the existing schedule to some extent for the jobs to be processed on that period. If a job is being processed on a certain machine when a disruption occurs to the machine, the job is suspended for a while. Let us express its remaining processing time by t. It is, in many cases, postulated that the suspended job is resumed after the prohibited period is over to be processed just for t. It is referred to as *resume*. However, it might not be a realistic assumption. It would be natural to postulate that the preempted job has to start its processing from the beginning when it is put back on the machine. This is called *restart*. Figure 3.1 depicts *resume* and *restart* of job processing against a machine breakdown.

In addition to the above, jobs initially allocated on the prohibited time period are moved to a substitutable resource if exists, or simply moved rightward along the time axis.

Interruption refers to uncertainty caused by external factors such as listed below:

- Urgent job arrival caused by a rush order occurring within the target planning horizon [28];
- Cancellation of scheduled jobs by which its succeeding jobs would be shifted forward [31, 37];
- Changes in job specification by which new tasks are to be added, or existing tasks are to be removed [31];
- Changes in due dates of a jobs, delays of material deliveries and pegged requirements (pegging) (e.g., [12, 14, 17]).

Variation refers to the uncertainty associated with jobs themselves. The unexpected events categorized into *variation* are basically originated within a planned job, and are of spontaneous and frequent occurrence (e.g., [4, 5, 7, 9]). Their impact on the progress of the existing schedule would be smaller than those categorized into *disruption* or *interruption*. In general, *variation* signifies the difference between the

Fig. 3.1 Machine breakdown model

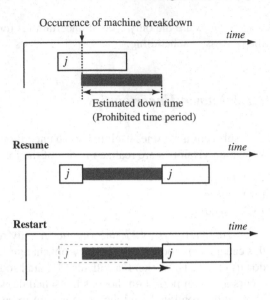

actual processing, setup or release time of a job and its initially planned one, or the extent or degree to which the difference varies. The variation of job processing times and setup times would be closely related to the worker's skill [18, 34].

The *known uncertainty* described above may delay the existing schedule in many situations, while there would occasionally be a case where processing of some jobs are terminated earlier than planned due to overestimation. In relation to *known uncertainty*, Wu et al. [36] have discussed a so-called *generalized approach* when a machine breakdown can be a considered as the only factor of known uncertainty. Huang et al. [16] also dealt with a similar problem which is referred to as a *machine breakdown problem*.

In their approaches, they considered a case where a certain machine, say μ_k, becomes down for some reasons. Let t_b denote the time at which the breakdown occurs, and t_r express the time when its recovery is completed. Then, the time λ spent in repairing the machine, μ_k is given by $\lambda = t_r - t_b$, which expresses the prohibited period of the machine on the schedule. If a job should be being processed at t_b on μ_k, its processing would be suspended. After μ_k recovers, processing of the job is to resume or restart on μ_k under the condition of preemption. However, their approaches have confined themselves only to the unexpected events belonging to *disruption*, ignoring interruption and variation. This is because all the factors categorized into *interruption* or *variation* do not always generate the prohibited time period on the schedule, and in addition even if some unexpected events in *interruption* or *variation* generate a prohibited time on the schedule, its starting time does not agree with the arrival time of an unexpected event. In the case of an urgent job, for example, its starting time does not necessarily agree with its arrival time since it would adaptively be involved in the series of scheduling decisions.

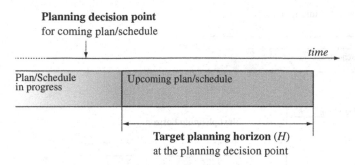

Fig. 3.2 Planning decision point and planning horizon

3.2 Planning Horizon and Related Matters

Planning horizon: A planning horizon, denoted by H, refers to a time span consisting of multiperiods [10, 23]. Over the planning horizon, a target plan or schedule is created on the basis of the outcome of its higher level decision-making unit, where a period represents a day, a week, a ten-day, or a month, depending on the firm's managerial policy or the category of products to be manufactured.

Planning decision point: Throughout this book, a point in time on which all the decision makings relating to planning or scheduling over a certain planning horizon are actually conducted is called *planning decision point*. At a planning decision points, the decision maker creates a capacity plan and a detailed shop floor schedule for the forthcoming planning horizon consisting of multiperiods. Then, the plan or schedule for the first period is released at the beginning of the first period. Figure 3.2 shows the relation between the planning decision point and its planning horizon.

Planning Cycle: As shown in Fig. 3.3, individual plans or schedules have an overlapped time with their consecutive ones in the real manufacturing environments. The interval of time between the beginnings of two consecutive plans is called a *planning cycle* and denoted by c since it is almost a constant.

The short-term capacity plan is, for instance, updated monthly, and the shop floor schedule is then updated weekly or daily on the basis of the short-term capacity plan. In this case, the planning cycle corresponds to that of the shop floor schedules.

Rolling Horizon Procedure: The regular updating of the current plan or the existing schedule by shifting the planning horizon a suitable amount toward the future is called *rolling horizon procedure* [25, 26] or *rolling schedules* [8, 19]. This is easy to understand by focusing on shop floor schedules.

As shown in Fig. 3.4, suppose that the planning horizon H_p for a short-term capacity plan consists of four periods, and that a short-term capacity plan with the rough schedule for the four periods from the $(t-1)$th to the $(t+2)$th period have been already created at the end of the $(t-2)$th period. Then, the decision maker

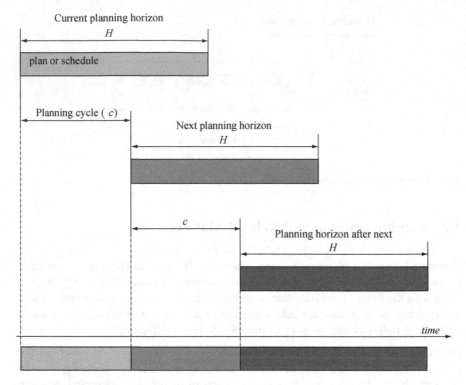

Fig. 3.3 Planning horizon and planning cycle

generates a detailed shop floor schedule for the first period, i.e., the $(t-1)$th period, where each period has a planning horizon H_s.

Rolling horizon procedure will, at the end of the $(t-1)$th period, update or revise the short-term capacity plan for the coming four periods, from the tth to the $(t+3)$th periods, taking into account new jobs arriving at the $(t-1)$th period. At this time, the rough schedules for the tth through the $(t+3)$th periods are generally obtained as by-products of this activity. And then, the detailed shop floor schedule is generated for the tth period.

The above is a rolling horizon procedure against the short-term capacity plan with a planning horizon H_p. In addition, Fig. 3.4 also shows that each individual period consists of three short periods, and that the similar approach to the rolling horizon procedure is applied to the planning horizon H_s, indicating a nesting structure.

After each detailed shop floor schedule is generated, the production instructions relating to the schedule are released into the shop floor, indicating the information described in the shop floor schedule, where the information is usually on job allocations to proper resources, task sequences and so forth. At this stage, releasing production instructions is called dispatching. This is why the shop floor schedule is occasionally referred to as the *dispatching schedule* in the real circumstances.

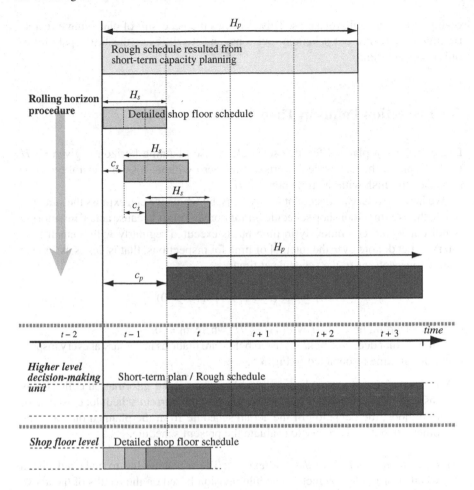

H_p Planning horizon of a short-term capacity plan and rough schedule
c_p Planning cycle of short-term capacity planning

H_s Planning horizon of a shop floor schedule
c_s Planning cycle of off-line scheduling

Fig. 3.4 Updates of the shop floor schedule

Online Scheduling: During the process of manufacturing, the shop-floor schedule shall undergo some fluctuations in various attributes due to uncertainties, and these fluctuations will reduce productivity of the manufacturing system because they will disturb the smooth progress of the planned shop floor schedule. With a view to coping with this problem and keeping the feasibility of the current schedule, schedule revisions is introduced as necessary. However, frequent schedule revisions will also disturb the smooth progress since the manufacturing process is forced to discontinue owing to the schedule revisions. This indicates the necessity of clarifying effectual timings of carrying out schedule revisions, and it will exert the full capability of online scheduling.

3.3 Inspection Points in Time

Let, S denotes a planned feasible schedule whose planning horizon is given by H (>0). Suppose that schedule S starts at time zero without loss of generality and is supposed to finish with the time horizon H.

We here introduce a concept of *inspections*, by which we can express the situation where the existing floor shop schedule is monitored. This is because actual monitoring is not conducted continuously in time but is executed regularly with a small time interval. Let denote by τ the interval of time for inspections, that is, let us assume to conduct inspections to the schedule at times

$$\tau_i = i\tau \ (i = 1, \ldots, M, \tau > 0)$$

over the planning horizon H with $\tau_M \leq H$, where $\tau_0 = 0$.

Suppose that decision makings listed below are made as necessary at every inspection point in time as depicted in Fig. 3.5:

(1) *Review of the current schedule*: Based on aggregated information collected by inspection, the scheduler reviews progress of the current schedule: existence of tardy jobs, the deviation of the actual schedule from the planned one, and the other attributes necessary to evaluate the present schedule.

(2) *Decision to conduct a schedule revision*: The scheduler has to make a decision whether or not to conduct a schedule revision based on the results of the above review. Once it is decided to conduct a schedule revision, the present inspection point in time concurrently plays a role of a *scheduling point*, and the next decision described in (3) has to be made. Otherwise, the current schedule is carried on without any modification to the job sequence in the schedule or so-called right-shift operations are applied to the existing schedule to keep its feasibility.

(3) *Execution of revision*: It takes time to conduct a schedule revision against the present schedule, and the revised schedule is not released immediately after it is obtained. As shown in Fig. 3.6, the newly generated schedule becomes available at its release time.

Some suitable criteria are necessary to make a judgment on whether or not to execute a schedule revision at each inspection point. In this book, we introduce two kinds of criteria for this judgment; one is a *periodic* basis and the other is an *event-driven* one.

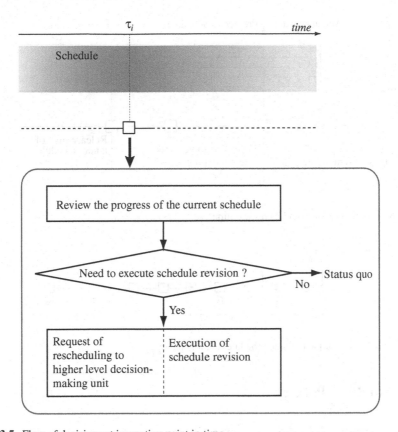

Fig. 3.5 Flow of decisions at inspection point in time

The periodic basis is one of the simplest ones, and substantially the same as the underlying idea for rolling schedules defined in Sect. 3.1. It urges us to conduct a schedule revision at times $T, 2T, \ldots$, where $T = l\tau$ and l is a positive integer. The schedule revision policy on the periodic basis is called a *periodic schedule revision policy* in the following.

Upon the event-driven basis, on the other hand, the schedule revisions are triggered by designations by higher level decision makers or operators due to rush orders, machine breakdowns, and the other unexpected events. It is called an *event-driven schedule revision policy*.

It is also possible to combine the above two policies. This book deals with two more schedule revision policies, which can be devised by combining the two bases: one is called a *hybrid policy* and the other is an *enhanced event-driven policy*. Under the hybrid policy, the schedule is revised at times $T, 2T, \ldots$, and also triggered by some specific events. Under the enhanced event-driven policy, on the other hand, the present schedule is revised when the elapsed time since the previous schedule revision becomes T or some specific event occurs whichever occurs first.

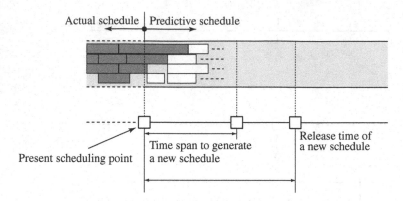

Fig. 3.6 Schedule revision and its release time

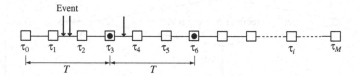

Fig. 3.7 Behaviors of the periodic schedule revision policy

3.4 Two Basic Policies

3.4.1 Periodic Schedule Revision Policy

Under the periodic schedule revision policy, the current schedule is revised at time $T, 2T, \ldots$, where T is given by

$$T = l\tau, \tag{3.1}$$

and τ signifies the interval of two successive inspections with a positive integer l. Figure 3.7 shows behaviors of the periodic schedule revision policy. In this figure, black dots in the squares on the time axis indicate that a schedule revision was carried out there, and that the inspection interval is $T = 3\tau$. Under this policy, the design variable is T or l.

This policy has an advantage of managerial simplicity, but at the same time it has a disadvantage of being unable to cope with emergencies such as machine breakdowns and arrivals of urgent jobs promptly. Next subsection discusses the event-driven schedule revision policy which can overcome such a disadvantage.

Fig. 3.8 Behaviors of the event-driven schedule revision policy

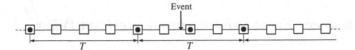

Fig. 3.9 Behaviors of the hybrid schedule revision policy

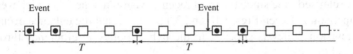

Fig. 3.10 Inefficient behaviors of the hybrid schedule revision policy

3.4.2 Event-Driven Schedule Revision Policy

Figure 3.8 shows behaviors of the event-driven schedule revision policy. It is observed in this figure that the event-driven schedule revision policy can deal with emergencies as promptly as possible at the expense of managerial simplicity compared with the periodic policy.

3.5 Advanced Policies

3.5.1 Hybrid Schedule Revision Policy

In the real manufacturing environments, schedule revisions are performed on a periodic basis, but an event-driven scheduling is also invoked by emergent events in many cases. It is called a *hybrid schedule revision policy*, having the advantages of both the periodic policy and the event-driven one. Figure 3.9 shows behaviors of the hybrid schedule revision policy. It is seen in this figure that every emergency can be dealt with dynamically and promptly.

However, the hybrid policy occasionally shows inefficient behaviors. Figure 3.10 reveals a case where schedule revisions are executed consecutively, e.g., once forced by the periodic basis and one more time due to an event, or once for an event and one more periodically. The enhanced event-driven schedule revision policy is discussed in the next subsection with a view to overcoming this inefficiency problem.

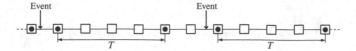

Fig. 3.11 Enhanced event-driven schedule revision policy

3.5.2 Enhanced Event-Driven Schedule Revision Policy

Figure 3.11 shows behaviors of the enhanced event-driven schedule revision policy. This policy conducts a schedule revision immediately when an emergent event occurs or when the elapsed time since the most recent revision reaches T, whichever occurs first. Comparison between Figs. 3.10 and 3.11 reveals that the enhance event-driven schedule revision policy is superior to the hybrid policy in efficiency through the managerial elaboration.

3.6 Literature Overview

Early studies on periodic schedule revision policies provide a generic framework rolling schedules [6, 24]. Baker and Peterson have proposed an analytic model to evaluate the performance of rolling for production planning problems. Muhlemann et al. have investigated the effectiveness of a periodic approach in job shop environments with inaccurate job processing times and machine breakdowns. They shows the relationship between the performance of utilized heuristics and the rescheduling period. Ovacik and Uzsoy have provided several heuristics for rolling horizon procedure against a single machine and parallel machines with sequence dependent setups, where dynamic job arrivals are considered [25, 26]. Shafaei and Brunn have examined the performance of dispatching rules under the periodic schedule revision policy based on practical data in a dynamic job shop environment [29, 30]. Vieira et al. [35] have proposed analytical models which can estimate the performance of a single machine system under the periodic and event-driven schedule revision policies . They considered dynamic job arrivals and machine breakdowns as unforeseen events. Their model can evaluate the adequacy of the policy, it is, however, postulated that both the underlying distributions of random events and the parameters involved are completely known beforehand.

The applicability of the hybrid approach has been pointed out by a number of literatures [11, 24]. From the more practical and managerial point-of-view, Suwa and Sandoh [32] have concentrated on the cumulative task delays to express the status of a job shop schedule instead of individual emergencies since they can constantly be monitored at each inspection point in time. Suwa [33] has also discussed the hybrid schedule revision policy focusing on the cumulative delays above mentioned. He considered the single machine dynamic scheduling with urgent jobs and sequence dependent setup times and minimized the total setups as well as frequency of rescheduling to show good performances of the hybrid schedule revision policy in minimizing the total setup times with less frequency of rescheduling through computational experiments. These topics will be discussed later in this book.

References

1. Albers S, Schmidt G (2001) Scheduling with unexpected machine breakdowns. Discret Appl Math 110(2–3):85–99
2. Allahverdi A, Mittenthal J (1994) Two-machine ordered flowshop scheduling under random breakdowns. Math Comput Model 20(2):9–17
3. Aytug H, Layley M, McKay K, Mohan S, Uzsoy R (2005) Executing production schedules in the face of uncertainties: a review and some future directions. Eur J Oper Res 161:86–110
4. Balasubramanian J, Grossmann IE (2000) Scheduling to minimize expected completion time in flowshop plants with uncertain processing times. Comput Aided Chem Eng 8:79–84
5. Balasubramanian J, Grossmann IE (2002) A novel branch and bound algorithm for scheduling flowshop plants with uncertain processing times. Comput Chem Eng 26(1):41–57
6. Baker KR, Peterson DW (1979) An analytic framework for evaluating rolling schedules. Manag Sci 25(4):341–351
7. Beraldi P, Ghiani G, Guerriero E, Grieco A (2006) Scenario-based planning for lot-sizing and scheduling with uncertain processing times. Int J Prod Econ 101(1):140–149
8. Blackburn JD, Millen RA (1982) The impact of a rolling schedule in a multi-level MRP system. J Oper Manag 2(2):125–135
9. Bonfill A, Espuña A, Puigjaner L (2008) Proactive approach to address the uncertainty in short-term scheduling. Comput Chem Eng 32(8):1689–1706
10. Chung C-H, Krajewski LJ (1984) Planning horizons for master production scheduling. J Oper Manag 4(4):389–406
11. Church LK, Uzsoy R (1992) Analysis of periodic and event-driven rescheduling policies in dynamic shops. Int J Comput Integr Manuf 5(3):153–163
12. Daniel Ng CT, Edwin Cheng TC, Kovalyov MY, Lam SS (2003) Single machine scheduling with a variable common due date and resource-dependent processing times. Comput Oper Res 30(8):1173–1185
13. Ecker KH, Gupta JND (2005) Scheduling tasks on a flexible manufacturing machine to minimize tool change delays. Eur J Oper Res 164(3):627–638
14. Geneste L, Grabot B, Letouzey A (2003) Scheduling uncertain orders in the customer-subcontractor context. Eur J Oper Res 147(2):297–311
15. Holthaus O (1999) Scheduling in job shops with machine breakdowns-an experimental study. Comput Ind Eng 36(1):137–162
16. Huang YG, Kanal LN, Tripath SK (1994) Reactive scheduling for a single machine: Problem definition, analysis, and heuristic solution. Comput Integr Manuf 3(1):6–12
17. Jia C (2001) Stochastic single machine scheduling with an exponentially distributed due date. Oper Res Lett 28(5):199–203
18. Kim SC, Bobrowski PM (1997) Scheduling jobs with uncertain setup times and sequence dependency. Omega 25(4):437–447
19. Kimms A (1998) Stability measures for rolling schedules with applications to capacity expansion planning, master production scheduling, and lot sizing. Omega 26(3):355–366
20. Li Z, Ierapetritou M (2009) Integration of planning and scheduling and consideration of uncertainty in process operations. Comput Aided Chem Eng 27:87–94
21. Liao CJ, Chen WJ (2004) Scheduling under machine breakdown in a continuous process industry. Comput Oper Res 31(3):415–428

22. McKay K, Buzacott J, Safayeni F (1989) The scheduler's knowledge of uncertainty: the missing link. In: Browne J (ed) Knowledge based production management systems: the IFIP WG5.7 working conference on knowledge based production management systems, Elsevier, Amsterdam
23. Millar HH (1998) The impact of rolling horizon planning on the cost of industrial fishing activity. Comput Oper Res 25(10):825–837
24. Muhlemann AP, Lockett AG, Farn CK (1982) Job shop scheduling heuristics and frequency of scheduling. Int J Prod Res 20(2):227–241
25. Ovacik IM, Uzsoy R (1994) Rolling horizon algorithms for a single-machine dynamic scheduling problem with sequence-dependent setup times. Int J Prod Res 32(6):1243–1263
26. Ovacik IM, Uzsoy R (1995) Rolling horizon procedures for dynamic parallel machine scheduling with sequence-dependent setup times. Int J Prod Res 33(11):3173–3192
27. Prickett PW, Johns C (1999) An overview of approaches to end milling tool monitoring. Int J Mach Tools Manuf 39(1):105–122
28. Raheja AS, Subramaniam V (2002) Int J Adv Manuf Technol 19(10):756–763
29. Shafaei R, Brunn P (1999) Workshop scheduling using practical (inaccurate) data Part 1: the performance of heuristic scheduling rules in a dynamic job shop environment using a rolling time horizon approach. Int J Prod Res 37(17):3913–3925
30. Shafaei R, Brunn P (1999) Workshop scheduling using practical (inaccurate) data Part 2: an investigation of the robustness of scheduling rules in a dynamic and stochastic environment. Int J Prod Res 37(18):4105–4117
31. Sun J, Xue D (2001) A dynamic reactive scheduling mechanism for responding to changes of production orders and manufacturing resources. Comput Ind 46(2):189–207
32. Suwa H (2007) A new when-to-schedule policy in online scheduling based on cumulative task delays. Int J Prod Econ 110:175–186
33. Suwa H, Sandoh H (2007) Capability of cumulative delay based reactive scheduling for job shops with machine breakdowns. Comput Ind Eng 53:63–78
34. Techawiboonwong A, Yenradee P, Das SK (2006) A master scheduling model with skilled and unskilled temporary workers. Int J Prod Econ 103(2):798–809
35. Vieira GE, Herrmann JW, Lin E (2000) Analytical models to predict the performance of a single-machine system under periodic and event-driven rescheduling strategies. Int J Prod Res 38(8):1899–1915
36. Wu SD, Storer RH, Chang PC (1993) One-machine rescheduling heuristics with efficiency and stability as criteria. Comput Oper Res 20(1):1–14
37. Zheng F, Chin FYL, Fung SPY, Poon CK, Xu Y (2006) A tight lower bound for job scheduling with cancellation. Inf Process Lett 97(1):1–3

Chapter 4
Methods for Online Scheduling

Abstract This chapter considers how schedules are generated and to be modified in online scheduling environments. We first categorize the online scheduling into two types, dispatching and schedule revisions. Second, we overview the procedure of dispatching and the well-known rules of scheduling, and then discuss the schedule revision to understand that a so-called right-shift operation and an iterative schedule revision are effective to cope with uncertainty caused by dynamic changes in manufacturing environments. We also glance at knowledge-based approaches which have played an important role in online scheduling.

4.1 Categories of Online Scheduling

Online scheduling consists of dispatching and schedule revisions as discussed in Sects. 1.2 and 2.2. The dispatching is to allocate a job to a machine one after another according to the prespecified dispatching rules when the machine becomes idle, while the schedule revisions are to modify or adjust the existing schedule so that it can accommodate the changes in manufacturing environments. We have two approaches for schedule revisions from the viewpoint of when-to-revise described in the previous chapter; one is to revise the existing schedule periodically, and the other is to revise the schedule every time an unexpected event occurs. The former approach is referred to as a *periodic schedule revision* and the latter as a *reactive scheduling*. Figure 4.1 depicts the positional relationship among the dispatching, the periodic schedule revision, and the reactive scheduling in the manufacturing execution phase.

When the dispatching is adopted, no specific predictive schedule is needed; dispatching rules will determine a job to be allocated next and this decision making is made locally in real time. Instead, we cannot guarantee the global optimality of the whole schedule over time, and it is hard to predict the performance and the progress of the manufacturing.

H. Suwa and H. Sandoh, *Online Scheduling in Manufacturing*,
DOI: 10.1007/978-1-4471-4561-5_4, © Springer-Verlag London 2013

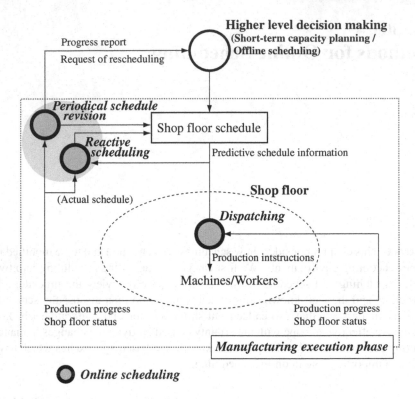

Fig. 4.1 Categories of an online scheduling

The well-known "simulation-based scheduling", which has been described in Chap. 1, is a practical extension of dispatching rules. In simulation-based scheduling, the currently used dispatching rule can adaptively be switched to another one, which is considered to be more suitable, in accordance with the changes in the manufacturing status and the environment. The rules for switching a dispatching rule to another are called *adaptive rules*. Section 4.2 briefly describes the procedure of dispatching, and Sect. 4.3 overviews the representative dispatching rules along with the adaptive rules.

On the other hand, schedule revisions are one of the essential decision makings in online scheduling to modify or adjust the current shop floor schedule so that it can accommodate the changes in the manufacturing environments appropriately. The basic procedure of schedule revision is described in Sect. 4.4. Policies for schedule revisions are largely identified into two categories; a periodical schedule revision policy and a reactive scheduling policy. The periodical schedule revision policy revises the existing schedule at prespecified times $T, 2T, \ldots$, whereas the reactive scheduling policy revises the existing schedule promptly every time an unexpected events occurs.

The periodical schedule revision policy is widely installed and easily managed because it underlies a simple structure of data relating to the manufacturing progress. Under the periodical schedule revision policy, however, disruptions in the demanding environments tend to invoke requests for rescheduling to the higher level decision-making unit. Once a schedule revision is determined to be performed, not only internal activities but external ones are aborted, and the rescheduling would be followed by significant changes in the details of the schedule.

Reactive scheduling is considered to be one of the most useful approaches to the problems that human schedulers should tackle in daily routine works for the purpose of ensuring the smooth progress of manufacturing as much as possible [8, 21]. This is because the reactive scheduling policy modifies or adjusts the existing schedule adaptively when some unexpected events occur. In other words, it is on an event-driven basis. This adaptivity is considered to be effective especially in demanding environments.

There exist the following two requirements in reactive scheduling:

- The firm planned schedule for a certain planning horizon generated steadily within the framework of deterministic or static scheduling to optimize some performance measure.
- A method for partial modification of the existing schedule rather than a rescheduling method that creates a new schedule from scratch. This implies that the quality of a predictive schedule generated at the planning phase should be as higher as possible.

Both requirements indicate that the quality of a predictive schedule will influence a result of reactive scheduling.

In adopting the reactive rescheduling, however, we should note the following:

- The initial predictive schedule for a certain planning horizon is generated assuming deterministic or static environments since the assumption reduces the intricacy of generating a secure schedule.
- Once the above initial schedule starts, partial modifications or adjustments to the schedule are more practicable than the rescheduling, creating a whole new schedule from scratch.

These circumstances in the manufacturing system indicate that the schedule will gradually and dynamically change itself, and hence the intermediate schedule as well as the initial one is simply a predictive schedule and is destined to change itself until it completes. For this reason, reactive scheduling is occasionally called a *predictive-reactive approach* [9, 23]. In contrast with the predictive–reactive approach, dispatching is referred to as *completely reactive approaches*.

Reactive scheduling can be also viewed within the framework of the scheduling-related decisions which aims at confronting with unexpected events by modifying or adjusting the affected part of the existing schedule. Most studies trying to deal with unexpected events have addressed the knowledge-based technology, and a representative example of the knowledge-based approach is outlined in Sect. 4.5.

4.2 Procedure of Dispatching

In the process of dispatching, priority indexes are assigned to the individual jobs which are waiting for being processed. The job with the highest priority is, then, picked up to be processed from the waiting jobs, or all the waiting jobs are rearranged in nondecreasing sequence of the priority index. The latter approach is generically called *sequencing*.

The common procedure of dispatching can be described as follows: Suppose that a single machine in a manufacturing stage is becoming available at time, denoted by t. In addition, let us consider that there exist, in the work-in-process goods (or loading buffer) of the stage, a set $N(t)$ of jobs which are to be processed on that machine.

Step 1 Compute the priority index $I_j(t)$ of job j in $N(t)$.
Step 2 Process the job with the highest priority among candidates.

Note that the above **Step 2** can be replaced by the following operation by means of the list scheduling method outlined in Sect. 2.4:

Step 2' The jobs in $N(t)$ are rearranged in nondecreasing order of priority indexes yielding an updated job list, or sequence.

Figure 4.2 depicts an illustrative example of dispatching at a certain stage with two machines R_1 and R_2. As shown in the upper part of Fig. 4.2, five jobs are work-in-process at time t, and thus we have $N(t) = \{3, 4, 5, 6, 7\}$. Suppose that machine R_1 is just about to complete the processing of job 1 at t, and that a single job is to be selected from among jobs in $N(t)$ according to some dispatching rule. The lower part of the figure shows that job 4 is assigned to R_1 yielding a locally updated schedule. In this way, we locally allocate the unprocessed jobs to a machine, which is available, in real time.

Restrictions on the information utilized in scheduling occasionally force dispatching rules to be myopic, accordingly the dispatching rules cannot necessarily sustain high system performance over a long time horizon [5]. To make things worse, the attributes inherent to myopia do not require detailed schedules at the planning phase, and hence it becomes significantly hard to predict the manufacturing progress and the system performance.

On the other hand, the rapid progress in the information and communication technologies have entirely changed the aspect of computer-aided scheduling. Aggressive studies based on computer simulation techniques have successfully accumulated useful and valuable knowledge on the characteristics of dispatching rules to develop an advanced approach in the scheduling environments, which is called simulation-based scheduling.

Simulation-based scheduling aims to realize the flexible decision-making under dynamic manufacturing environments with the help of computational simulations. The scheduling decisions in this approach is conducted by a sort of logic to change

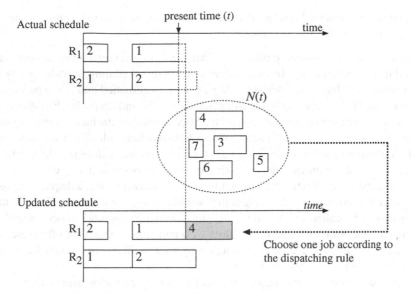

Fig. 4.2 Example of dispatching

over the dispatching rule to another along with environmental changes, and its underlying logics can be expressed by a form of adaptive rules. With the aid of technologies to accumulate useful and indispensable information, which is incorporated in the knowledge base, simulation-based scheduling has evolved into intelligent scheduling.

4.3 Scheduling Rules

4.3.1 Principle of Scheduling Rules

Throughout the scheduling decision process in the real circumstance, intricate and time-consuming methods, such as branch-and-bound or dynamic programming, are hard to be used due to procedural constraints as mentioned in Sect. 1.3. In particular, schedule revisions entail some efficient heuristic methods, which are composed of relatively simple scheduling operations, in allocating each individual job to a suitable machine within a limited time.

In online scheduling, the representative of such heuristics will be the so-called *scheduling rules*. The roles of these scheduling rules are:

- To provide logics to allocate priority indexes to individual jobs in the work-in-process inventory, and

- To enable a certain scheduling decision which is considered to be appropriate for the current manufacturing situation.

Among others, dispatching rules are widely known, and regarded as fundamental procedures for scheduling. In the academic fields of scheduling, a wide variety of dispatching rules have been devised. They are largely classified into two types, *static* and *dynamic* [13], according to the information on jobs and shops. Each dispatching rule of the static type involves a priority index computed on the basis of one or more job attributes which take on constant values. A dispatching rule of the dynamic type holds some more job attributes which are time-dependent, such as the slack time to the due date and the remaining processing time that may dynamically change.

An interesting application of dispatching rules is to switch over a dispatching rule to another in accordance with changes in a manufacturing environment. Rules in this approach will be called a *meta rule* since each rule in this approach is occasionally a composite of several different rules. Most such meta rules, which are often referred to as *adaptive rules*, have been developed by means of AI- or knowledge-based techniques as will be introduced later.

In the remainder of this chapter, we deal with the three scheduling rules; *static dispatching rules*, *dynamic dispatching rules*, and *adaptive rules*. The following notations will be used:

$N(t) =$ A set of waiting jobs at time t.
$I_j(t) =$ Priority index of job j in $N(t)$.
$r_j =$ Release date of job $j \in N(t)$.
$p_j =$ Processing time of job $j \in N(t)$.
$d_j =$ Due date of job $j \in N(t)$.
$w_j =$ Importance factor (or weight) of job $j \in N(t)$.
$v_j(t) =$ Remaining processing time of job $j \in N(t)$ $(0 \leq v_j(t) \leq p_j)$.

4.3.2 Static Dispatching Rules

Static dispatching rules are composed of *static attributes* such as p_j, r_j and d_j, which are constant and independent of the elapsed time. Their representatives that are considered to be significant tools for decision making in practice are as follows:

The Earliest Release Date (ERD) rule:

– The priority index, I_j, of job j is given by

$$I_j = 1/r_j. \tag{4.1}$$

The ERD rule defined above is equivalent to the First Come First Served (FCFS) rule or the First In First Out (FIFO) rule. In static scheduling decision makings, the ERD rule arranges the jobs in $N(t)$ in nondecreasing order of r_j. The ERD rule is also widely known in queueing and scheduling theories.

The Earliest Due Date (EDD) rule:

– The priority index, I_j, of job j is defined by

$$I_j = 1/d_j. \tag{4.2}$$

The jobs in $N(t)$ are sorted in a nondecreasing order of d_j. The EDD rule can provide an optimal solution to a single machine static scheduling problem where the maximum tardiness is minimized. There exist various transforms of the EDD rule, one of which is the so-called preemptive EDD rule. The preemptive EDD rule proves itself against single machine static scheduling problems with job release time and preemption where the objective is to minimize the maximum tardiness.

The Shortest Processing Time (SPT) rule:

– The priority index, I_j, of job j is defined by

$$I_j = 1/p_j. \tag{4.3}$$

The jobs in $N(t)$ are arranged in a nondecreasing order of processing time p_j. The SPT rule generates an optimal schedule for a single machine static scheduling problem which minimizes the total completion time. The Weighted Shortest Processing Time (WSPT) rule is a generalized version of the SPT rule, which arranges the jobs in $N(t)$ in nondecreasing order of p_j/w_j (i.e., $I_j = w_j/p_j$). The WSPT rule provides an optimal schedule for a single machine static scheduling problem where the objective function is the total weighted completion time to be minimized.

4.3.3 Dynamic Dispatching Rules

Dynamic dispatching rules utilize *dynamic attributes* of jobs that are dependent on time t in addition to static attributes such as processing times. Interestingly, the job priorities vary over time under the dynamic dispatching rules. A symbol, $I_j(t)$, of time, t, is employed to denote the priority of job j instead of I_j used in the static dispatching rules. The following are representatives of the dynamic dispatching rules.

The Minimum Slack Time (MST) rule:

– The priority of job j in $N(t)$, which is not overdue, is given by

$$I_j(t) = 1/\left(d_j - v_j(t) - t\right). \tag{4.4}$$

To an overdue job, the highest priority (e.g., $I_j(t) > 1$) is given in order to process it first.

The Critical Ratio (CR) rule:

- One may say that, in practice, a job is critical if its slack time is relatively small and its importance factor, w_j, is relatively large. In reflection of these ideas, the *critical ratio* of a job is given by $(d_j - t)/w_j$, and thereby the priority index $I_j(t)$ of job j $(d_j > t)$ is defined by

$$I_j(t) = w_j/(d_j - t). \tag{4.5}$$

The Modified Due Date (MDD) rule:
- The MDD rule is an extension of the EDD rule to accommodate dynamic changes in a manufacturing environment. Its priority index is given by

$$I_j(t) = 1/\max\left(d_j, t + v_j(t)\right). \tag{4.6}$$

The Apparent Tardiness Cost (ATC) rule:

- The ATC rule is considered to be a useful and powerful dispatching rule because it involves the advantageous characteristics of two practical rules, the WSPT and the MST rules. Its priority index based on the so-called tardiness cost is defined by

$$I_j(t) = \frac{w_j}{p_j} \exp\left(-\frac{\max(d_j - v_j(t) - t, 0)}{\kappa \bar{p}}\right), \tag{4.7}$$

where κ and $\bar{p}$ respectively represent the scaling parameter and the average processing time of the remaining jobs. When κ is considerably small, the ATC rule shows its performances just like the MST rule against non-tardy jobs in $N(t)$, and like the WSPT rule for the tardy jobs. On the contrary, when κ takes on a large value, the ATC rule reduces to the WSPT rule.

4.3.4 Adaptive Rules

The adaptive rule, often referred to as a *production rule* in the field of Artificial Intelligence, is expressed by the following simple form:

$$STATUS \Rightarrow CLASS,$$

where *STATUS* consists of a conjunction of predetermined attributes while *CLASS* signifies an evaluation of *STATUS* with a discrete representation (e.g., *good* or *bad*), or a scheduling operation itself. Furthermore, the operator
$\Rightarrow$ is used in a fashion as

    (*Manufacturing status at a given time*)
    $\Rightarrow$ (*Scheduling operation considered to be effective at the time*)

Fig. 4.3 Example of adaptive rules by a decision tree representation

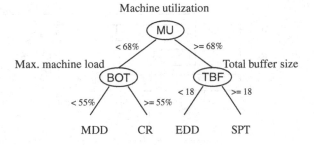

The manufacturing status corresponding to *STATUS* can be described by a sequence of values of job attributes and/or system attributes such as the amount of work-in-process inventory, machine utilization, and instantaneous machine load. As a scheduling operation, *CLASS*, we can specify the dispatching rules introduced above, or a certain priority of a given dispatching rule [7, 19].

A typical representation of the adaptive rule given by a so-called *if-then* rule, has been used in knowledge-based scheduling systems incorporating AI techniques. The other expression of the adaptive rule includes a *case*, or a *decision tree* [10, 14, 19]. Figure 4.3 depicts an example of a decision tree. Each node of the decision tree represents a predetermined attribute for the manufacturing status (*STATUS*). The decision tree in Fig. 4.3 explains which dispatching rule we should apply to a specific status. Following a path from the root node (MU) to the terminal node, we can extract four if-then rules as follows:

if MU < 68 and BOT < 55 then MDD
if MU < 68 and BOT ≥ 55 then CR
if MU ≥ 68 and TBF < 18 then EDD
if MU ≥ 68 and TBF ≥ 18 then SPT

The first rule, for example, indicates that the MDD rule should be selected if the current machine utilization drops below 68(%) and the maximum machine load becomes lower than 55(%).

4.4 Principle of Schedule Revision

4.4.1 Basic Procedure

In executing a schedule revision policy, we need to prespecify the procedure:

(1) To sustain the feasibility of the existing schedule, and
(2) To determine jobs to be rescheduled.

Hopefully, a procedure to retain the feasibility can also restrain the deviation from the existing schedule and thereby increase of cost associated with setup times, reassignment of workforces, and so forth. The simplest way to attain this is to conduct

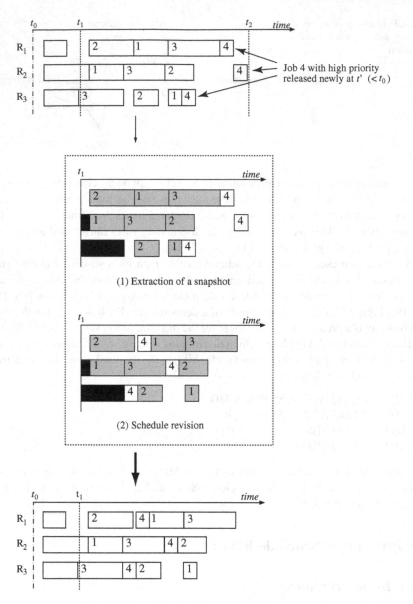

Fig. 4.4 Schedule revision by means of a snapshot

a right-shift operation [22]. It can generate a feasible schedule simply by shifting jobs, which have not been started, to the right as needed. The right-shift operation is described in detail in Sect. 4.4.2.

As for a procedure to select jobs to be rescheduled, a concept of time window is known to be useful. When an unforeseeable event occurs at time, t', during the manufacturing phase, a schedule revision is applied to the jobs within a time window with its beginning time, t_1, and terminating time, t_2, where $t' < t_1 < t_2$. The partial schedule extracted by the time window, $[t_1, t_2]$, is called a *snapshot*. A scheduling problem in the snapshot can be solved within the framework of static and deterministic scheduling.

Figure 4.4 illustrates scheduling in the snapshot with three machines under a dynamic job shop environment [4]. It is supposed, in Fig. 4.4, that a revision of the current schedule is about to be performed at time, t_0, due to arrival of job 4 with its technological order $R_3 \rightarrow R_1 \rightarrow R_2$. The snapshot is extracted in the current schedule after determining the time window $[t_1, t_2]$ (Fig. 4.4(1)). Figure 4.4(2) depicts job 4 is removed to the left. The beginning time, t_1, of the time window is possibly determined considering the status of ongoing manufacturing as well as the time to spend on scheduling simulation and generation of a modified schedule in the snapshot. On the contrary, there is no decisive way to determine the ending time, t_2, of the time window mainly because the time window approach does not guarantee the acquirement of better alternatives; one may have to apply a try-and-error method to determine t_2. To overcome this nuisance, the *match-up* strategy [1–3] is known to be useful. The match-up strategy modifies the current schedule partially so that we can match up with the original schedule at some future point in time, which is referred to as a match-up point. We can substitute the match-up point in time for the terminating time, t_2.

Figure 4.5 shows a typical process of schedule revision. This figure assumes a situation under which the processing of job 1 on R_2 was found to be delayed at time, t; this delay affects the starting time of jobs 1 and 2 on the same machine, R_2. The delay of job 1 on R_2 also invokes delays of other jobs which have not been started, and consequently causes the deviation of the actual schedule from the planned one.

At this stage, we should make a judgment whether or not a schedule revision is necessary. Once we determine to revise the schedule, a schedule revision is actually performed in the near future ("Revised schedule" in Fig. 4.5). Note that, when the delay is momentous, rescheduling is supposed to be requested to the higher level decision-making unit. On the contrary, if it is determined not to revise the schedule, job sequences on machines are not changed but the right-shift operation is applied to keep feasibility of the schedule ("Right-shift schedule" in Fig. 4.5).

4.4.2 Right-Shift Operation

In online scheduling, the *right-shift operation* is often applied to the existing schedule, which is challenged by a delay, in order to maintain the feasibility of the schedule [22]. The resulting schedule by applying the right-shift operation is called a *right-shifted schedule* which agrees with the original schedule in regard to the sequence of tasks and jobs.

Figure 4.6 shows an illustrative example of a right-shifted schedule under a job shop configuration with three machines and four jobs. At the present time, t, in

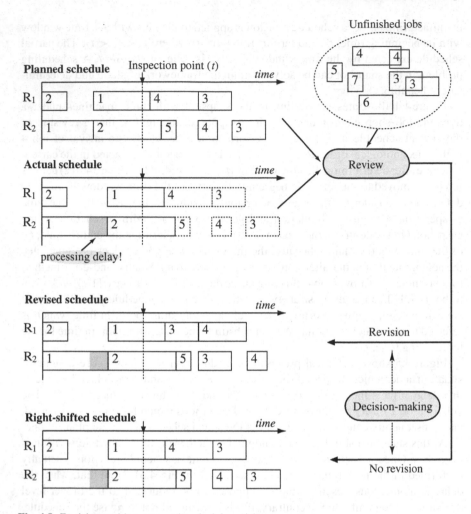

Fig. 4.5 Decision-making process in schedule revision

Fig. 4.6, machine R_2 is about to be down before starting job 1 on R_2 due to some trouble, e.g., a processing delay of Job 2 on R_2 or a setup delay of job 1. The predicted downtime which can be obtained at t is illustrated by the horizontal dark bar.

Under the above situation, the jobs on R_2 are focused on, and then jobs 1, 4, and 3 on R_2 are pushed to the right until R_2 becomes available. In accordance with this right-shift operation, these three jobs are shifted to the right on R_1 and R_3 if needed to maintain the feasibility of the current schedule. The lower chart in Fig. 4.6 indicates the resulting schedule. The right-shifted schedule is considered to be efficient since its deviation from the original schedule is relatively small, and moreover, it is recognizable for a human scheduler as well as workers on the shop floor who are confronting with schedule delays. In addition, the right-shifted schedule will effec-

Original schedule

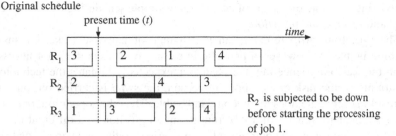

R₂ is subjected to be down
before starting the processing
of job 1.

Right-shifted schedule

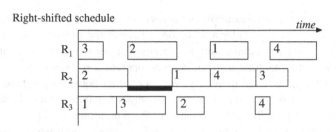

Fig. 4.6 Right-shift operation

tively refrain the cost increase associated with changes in setup times, reassignment
of resources including workforces, consistency of production information between
production control systems and machining systems with the built-in CAM software,
and the other complicated issues that are possibly difficult to quantify.

In the above, we have described the right-shift operation as if it were, in itself,
a method for schedule revisions. There are some who will say that the right-shift
operation itself is not a method for schedule revisions since they do not change the
sequence of jobs to be processed and there is some truth in this perspective if we
distinguish between *sequencing* and scheduling literally. In this book, however, the
right-shift operation is construed as a phase of a schedule revision, a schedule revision
occasionally consists of a single phase depending on circumstances though.

4.4.3 Iterative Revision

The next step after the right-shift operation in the process of schedule revision is to
find a better schedule by modifying the right-shifted schedule, which is expected to
reduce the uncertainties or the related costs. If some better schedules are successfully
found, one of them is released to the shop floor. Especially in the predictive-reactive
approaches, *iterative revision* (or *iterative repair*) [8, 10] has been widely studied as
a useful way of schedule revision. The iterative revision is a stepwise improvement of

the existing schedule by means of considerably simple scheduling operations called *swap* and *remove and insertion*.

The operation *swap* refers to an interchange of pairwise tasks. On multiple machine problems, however, a pairwise interchange of tasks does not necessarily retain the feasibility since the interchange unexpectedly violates the technological constraints on the task orders. For this reason, the swap is carried out only when the resulting schedule does not become infeasible. The other operation, *remove and insertion*, refer to transfer of a job from its initial position to a different one. This operation is slightly more complicated in comparison to the *swap* operation because the former operation necessitates both selection of a job to remove and determination of a position to insert it.

Figure 4.7 shows an illustrative example of the revision process by means of swap operations. In Fig. 4.7, let us consider a situation, where at the present inspection point in time, t, job 1 on R_2 cannot start as planned for some reason, but job 4 can start at t on the same machine instead. A swap operation between jobs 1 and 4 has, therefore, been applied, and thereby yielded a conflict between two tasks of job 1; the task of job 1 on R_1 and that on R_2. This conflict has induced a right-shift operation targeting at job 1 on R_1, and then job 4 on R_2 to maintain the feasibility of the schedule. It should be noted here that if the schedule is to be optimized, a swap operation is, moreover, necessary between jobs 1 and 4 on R_1.

Noronha [12] and Shaw [19] have, respectively, developed a heuristics and a knowledge-based method for iterative revisions. Among others, useful are heuristics based on a local search procedure such as tabu search, simulated annealing and genetic algorithm [11, 16]. These approaches, generically referred to as *metaheuristics*, seek a better schedule than the current schedule in *neighborhood* of the current schedule, and the search is carried out iteratively until a satisfactory schedule is obtained. The neighborhood consists of schedules, called neighbor schedules, which can be obtained by partially modifying the current schedule through simple operations such as swap or remove and insertion. In the context of revision, we start out with a right-shifted schedule and then evaluate each individual candidate schedule in its neighborhood based on some suitable criterion to detect a better schedule than the present one. Once better schedules are detected, we move to the best one among them to explore a further better schedule. Iterative revisions in this manner are conducted until a satisfactory schedule is obtained.

The knowledge-based approach has also played an important role in the development of online scheduling. The next section describes the details of this approach.

4.5 Knowledge-Based Reactive Scheduling

Over the last three decades, efforts have been devoted to develop reactive scheduling systems especially in the field of Artificial Intelligence. The conceptual framework of the reactive scheduling was studied as an application of the knowledge-based system in the 1980s [12]. From a practical point of view, however, it has a problem

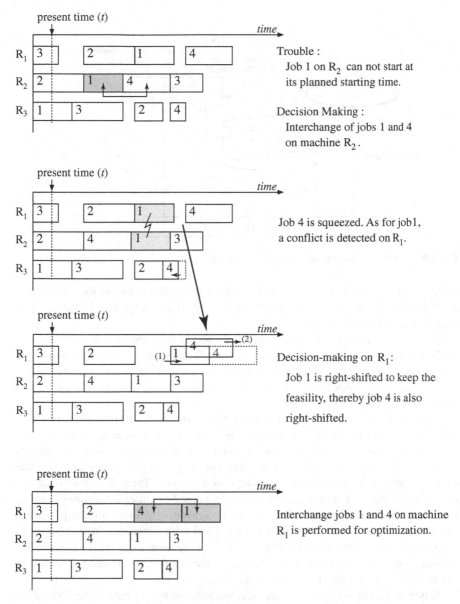

Fig. 4.7 Decision-making process with swap operations

we cannot overlook. First of all, the information extracted from experts are initially ambiguous in many cases, and therefore it should be transformed into some explicit and stereotypical form such as cases, decision trees, or if-then rules so that it is easily organized. This process of extracting information from experts and transforming it into a suitable form is called knowledge acquisition. Secondly, the extracted knowledge is not always effective in the changeable environment of the manufacturing

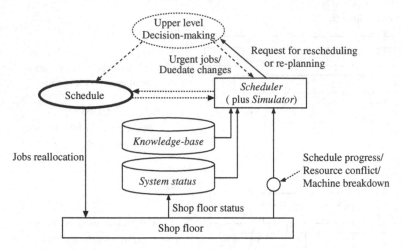

Fig. 4.8 Knowledge-based reactive scheduling system

system, and it is hard to append new useful knowledge dynamically even if it is available. For these reasons, the development of the knowledge-based system faded away for a period of time.

In these two decades, however, the technologies concerning information accumulation and data mining have shown remarkable advances, and concurrently the technological transfer has become a significant problem in manufacturing. These environmental changes have boosted the knowledge-based reactive scheduling and its system development.

Figure 4.8 shows a general framework of knowledge-based reactive scheduling systems [19]. At first, the shop floor gives an interrupt signal to request for a schedule revision because of a tool breakage, a resource conflict, or a trouble brought by workers. The higher level decision-making unit may also give a similar signal due to urgent jobs, job cancellation, or change of due dates. Then, *Scheduler* makes a decision whether or not to revise the schedule. Once *Scheduler* determines to revise the schedule, it attempts to revise the existing schedule by means of computational simulations. This is why the system is equipped with *Simulator*, which performs computational simulations reflecting the state of the shop floor on the basis of the information in *System status*. The outcomes of the simulations are used to determine the conclusive procedure to revise the schedule.

A typical example of the practical reactive scheduling system is CABINS developed by Miyashita and Sycara [10], which employs iterative revisions very efficiently to realize reactive scheduling. The quintessence of this system is the introduction of *case-based reasoning*, which is an artificial intelligence paradigm for reasoning and learning. CABINS is equipped with the following three functions:

- Construction of a predictive detailed schedule.
- Manual schedule revision by means of interaction with a human scheduler along with construction of case base.

- Automatic schedule revision by case-based reasoning.

A schedule revision is conducted by repetitively changing the starting time of a focal task, which is selected by some criterion at a certain time. The operation to change the starting time of a job is referred to as *shift*.

The case base is constructed in the following fashion: First, the system provides several preferable solutions to a human scheduler for a certain status of the manufacturing system, where the solutions consist of a series of shift operations to the focal task. The human scheduler selects one of them at his/her own discretion; empirically for instance. Second, a case is generated by combining the status and the solution, and the results are stored in the case base.

CABINS has basically six revision operations for focal tasks:

- *left_slide*: Shift the focal task to the left as much as possible within its repair time without changing its resource and the processing order of tasks on that resource.
- *left_shift*: Move the focal task leftward (Insert the focal task between two consecutive tasks to be processed prior to the focal task on the same machine) as much as possible within its repair time without changing the machine.
- *left_shift_alt*: Move the focal task leftward as much as possible on the substitutable resource that can process the focal task within its repair time.
- *swap*: Interchange the focal task with a certain task that is to be processed prior to the focal task on the same machine within the repair time.
- *swap_alt*: Interchange the focal task with some suitable task by allowing the focal task to move to the substitutable resource.
- *give_up*: Abandon all the revision above with respect to the focal task.

A series of one or more operations forms a procedure of schedule revision. The shifted schedule is created for the succeeding tasks of the focal along with the schedule revision procedure for the focal task.

In CABINS, the case-based reasoning (CBR) is utilized to select the best procedure of schedule revision. After applying a revision to the focal task, the CBR also reviews its outcome. If the result of the review is not satisfactory, another revision procedure is applied to the focal task. Otherwise, the system attempts to find out another task to be revised.

4.6 Literature Overview

Rajendran et al. [15] have investigated the effectiveness of a great number of dispatching rules, including those treated here, in flow shop systems as well as job shop systems. Shafaei et al. [17, 18] have examined the performance of both static and dynamic dispatching rules under dynamic job shop environments. They demonstrate that combination of the SPT rule and the CR rule is effective for real-time decision-making. Jeong et al. [7] have proposed a real-time scheduling system in which 16 dispatching rules are implemented. This system performs relatively small-scaled simulation before the dispatching process, and then the scheduler unit selects a dispatching rule to be applied to the current situation based on the simulation results.

Some works have attempted to apply metaheuristics to schedule revisions. Biewirth et al., for example, proposed a schedule repair method based on genetic algorithm [4], assuming a situation in which release times of jobs are not deterministic; jobs arrive randomly in real time at the manufacturing execution phase. They also investigated methods for schedule improvements under dynamic environments to propose to repair the schedule by applying genetic algorithm based on the snapshot approach as shown in Fig. 4.4.

Huang et al. [6] have proposed an analytical method for reactive scheduling in single machine scheduling problems where the maximum tardiness is to be minimized. The proposed method of iterative repair stands on job interchanges. They also introduced a switching point which is a point in time to apply the interchange operations after a machine breakdown occurred. The switching point can be obtained in the form of a function associated with job tardiness with respect to elapsed time.

An intelligent scheduling system, which automatically acquires heuristics, has been developed [20]. This system uses an inductive learning method to acquire heuristic rules. The suitability of each acquired scheduling rule is individually evaluated by simulation, and its preferable manufacturing STATUS, which was observed in Sect. 4.3.4, is obtained.

References

1. Aktūrk MS, Gūrel S (1999) Match-up scheduling under a machine breakdown. Eur J Oper Res 112(1):81–97
2. Aktūrk MS, Atamturk A, Gūrcl S (2010) Parallel machine match-up scheduling with manufacturing cost considerations. J Sched 13:95–110
3. Bean JC, Birge JR, Mittenthal J, Noon CE (1991) Matchup scheduling with multiple resources, release dates and disruptions. Oper Res 39(3):470–483
4. Bierwirth C, Mattfeld DC (1999) Production scheduling and rescheduling with genetic algorithm. Evol Comput 7(1):1–17
5. Church LK, Uzsoy R (1992) Analysis of periodic and event-driven rescheduling policies in dynamic shops. Int J Comput Integr Manuf 5(3):153–163
6. Huang YG, Kanal LN, Tripath SK (1990) Reactive scheduling for a single machine: Problem definition, analysis, and heuristic solution. Comput Integr Manuf 3(1):6–12
7. Jeong KC, Kim YD (1998) A real-time scheduling mechanism for a flexible manufacturing system: using simulation and dispatching rules. Int J Prod Res 36(9):2609–2626
8. Kerr R, Szelke E (1995) Artificial intelligence in reactive scheduling. Chapman and Hall, London
9. Mehta SV, Uzsoy RM (1998) Predictable scheduling of a job shop subject to breakdowns. IEEE Trans Robotics Autom 14(3):365–378
10. Miyashita K, Sycara K (1995) CABINS: a framework of knowledge acquisition and iterative revision for schedule improvement and reactive repair. Artif Intell 76:377–426
11. Morton TE, Pentico DW (1993) Heuristic scheduling systems. Wiley, New York
12. Noronha SJ, Sarma VVS (1991) Knowledge-based approaches for scheduling problems: a survey. IEEE Trans Knowl Data Eng 3(2):160–171

13. Pinedo M (2008) Scheduling - theory, algorithms, and systems, 3rd edn. Springer, New York
14. Quinlan JR (1993) C4.5: Programs for machine learning. Morgan Kaufman, San Mateo
15. Rajendran C, Holthaus O (1999) A comparative study of dispatching rules in dynamic flowshops and jobshops. Eur J Oper Res 116:156–170
16. Resende MGC, Sousa JP (2004) Metaheuristics: computer decision making. Kluwer Academic Publishers, London
17. Shafaei R, Brunn P (1999) Workshop scheduling using practical (inaccurate) data . Part 1: The performance of heuristic scheduling rules in a dynamic job shop environment using a rolling time horizon approach. Int J Prod Res 37(17):3913–3925
18. Shafaei R, Brunn P (1999) Workshop scheduling using practical (inaccurate) data. Part 2: An investigation of the robustness of scheduling rules in a dynamic and stochastic environment. Int J Prod Res 37(18):4105–4117
19. Shaw MJ, Park S, Raman N (1992) Intelligent scheduling with machine learning capabilities - the induction of scheduling knowledge. IIE Trans 24(2):156–168
20. Shaw MJ (1998) Introduction to the special issue on information-based manufacturing. Int J Flex Manuf Syst 10:195–196
21. Smith SF (1995) Reactive scheduling systems. In: Brown DE, Scherer WT (eds) Intelligent scheduling systems. Kluwer Academic Publishers, Boston, pp 155–192
22. Wu SD, Storer RH, Chang PC (1993) One-machine rescheduling heuristics with efficiency and stability as criteria. Comput Oper Res 20(1):1–14
23. Yang B, Geunes J (2008) Predictive-reactive scheduling on a single resource with uncertain future jobs. Eur J Oper Res 189:1267–1283

References

Chapter 5
Cumulative Delay-Based Schedule Revision Policy

Abstract This chapter describes a new approach to online scheduling, which focuses on the cumulative delays of the schedule as a criterion for the purpose of making a decision whether or not a schedule revision is invoked. This is because the cumulative delays can be considered as consolidated information to see if the current schedule is making smooth progress, and further they can be collected constantly without much effort at individual inspection points in time. This chapter describes the concept and definition of cumulative delays first and discusses a schedule revision policy based on them.

5.1 Cumulative Delay

Chapter 3 discussed the details of *known uncertainties*, which cause a variety of events, by classifying them into three categories; disruption, interruption, and variation. From the managerial point of view, however, if there exists a simple index that is easy to deal with for the purpose of measuring the progress of the current schedule that would be better.

The cumulative delay on a schedule can be viewed as consolidated information associated with uncertainty. This section concentrates on the cumulative delays of individual tasks of the schedule, and then designates the total cumulative delays to serve as the indicator to grasp the progress of the schedule.

5.1.1 Actual Delay of Each Job/Task

This subsection discusses the concept of an actual delay of each individual job or task. Let S denote a predictive schedule obtained at the planning phase. Manufacturing proceeds on the shop floor according to S, and the initial predictive schedule S would

H. Suwa and H. Sandoh, *Online Scheduling in Manufacturing*,
DOI: 10.1007/978-1-4471-4561-5_5, © Springer-Verlag London 2013

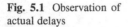
Fig. 5.1 Observation of
actual delays

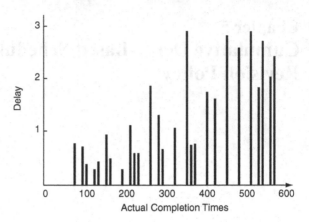

undergo some revisions as the occasion arises. At the end of the planning horizon of
S, therefore, the actual schedule, denoted by $\tilde{S}$, would be different from the planned
schedule S. This indicates that, at this stage, we can observe an actual delay of each
individual job. When job j is completed later than its planned time, there exists an
actual delay, denoted by $D(j)$ for job j. In addition, the delay may violate the due
date of job j, and what is worse, the delay may affect some external activities that
have been projected in accordance with S.

The *actual delay* $D(j)$ *of job* j is given by

$$D(j) = \max\{C_j(\tilde{S}) - C_j(S), 0\}. \tag{5.1}$$

In the case of $D(j) = 0$, it means that job j has completed as planned or before
the planned completion time. The actual job delays, if they are accumulated, tend
to show lower performance of the schedule, e.g., by extending the total completion
time or by enlarging the maximum tardiness since they are nondecreasing in the
completion time of individual jobs.

Likewise, we can deal with actual delays for individual tasks, that is, the *actual
delay* $D(j, k)$ *in task* ϕ_{jk} of job j on machine k can be expressed by

$$D(j, k) = \max\{C_{jk}(\tilde{S}) - C_{jk}(S), 0\}. \tag{5.2}$$

When it is revealed that $D(j, k)$ (>0) should be relatively major, the starting times
of its succeeding tasks might be delayed.

Figure 5.1 shows a distribution of actual delays of jobs observed on an illustrative
schedule, revealing that the actual delays tend to increase with the elapsed time.
However, it is generally difficult to clarify the details in relation to how the actual
delay of each individual job influences upon the performance measure such as the
total completion time or the maximum tardiness. For this reason, this chapter employs
the cumulative delay which is the aggregate amount of delays in completion times
of individual jobs or tasks.

Figure 5.2 reveals that when delays of jobs or tasks in Fig. 5.1 are accumulated, the resulting cumulative delay shows a curve which is increasing in the elapsed time at an accelerating rate. This suggests the importance of managing the actual delays of jobs or tasks in an integrated fashion.

5.2 Definition of Cumulative Delay

The concept of the cumulative delay described in the above is somewhat abstract. This subsection defines it more concretely to avoid its ambiguity by introducing two types of the cumulative delay which are commonly utilized through the remainder of this book.

5.2.1 Counting Up Only Completed Jobs or Tasks

Let H and S_0 denote a planning horizon and a planned schedule starting at time zero. As described in Sect. 3.3, inspections are conducted at inspection points in time $\tau_i = i\tau$ where $i = 1, 2, \ldots, M$ with a view to detect delays of job and to make a judgment whether to execute a schedule revision. It was also appended that these inspections play a role of monitoring the progress of manufacturing.

For an inspection point in time τ_i $(i = 1, 2, \ldots, M)$, let us denote by, $\tau_{[i]}$ $(\leq \tau_{i-1})$, the inspection time where a schedule revision was actually executed most recently prior to τ_i, and let $S_{[i]}$ express the resulting schedule from the schedule revision at $\tau_{[i]}$. It should be noted that $S_{[i]}$ plays a role of a predictive schedule at $\tau_{[i]}$ and that, if $\tau_{[i]} = l\tau$ for $l < i$, we have

$$\tau_{[l+1]} = \tau_{[l+2]} = \cdots = \tau_{[i-1]} = \tau_{[i]}.$$

In addition, if we should experience no schedule revision over $(0, \tau_i]$, we would have $\tau_i = 0$.

Even after the schedule revision, the revised schedule will undergo disruptions, interruptions, or variations which were discussed in Sect. 3.1. This signifies that the actual schedule realized over the period $(\tau_{[i]}, \tau_i]$ might slightly be different from the part of $S_{[i]}$ over the same period. For this reason, let S_i^A denote the part of the actual schedule over the period $(\tau_{i-1}, \tau_i]$, and let S_i^P express the corresponding part of $S_{[i]}$ to S_i^A over the same period.

Based on the above preliminaries, a cumulative delay of tasks can be defined as follows: let $\phi_{j_1 k}, \phi_{j_2 k}, \ldots, \phi_{j_\Lambda k}$, represent a sequence of tasks on machine k processed over the period $(\tau_{i-1}, \tau_i]$, and let *delay* of each task ϕ_{jk} $(j = j_1, \ldots, j_\Lambda)$

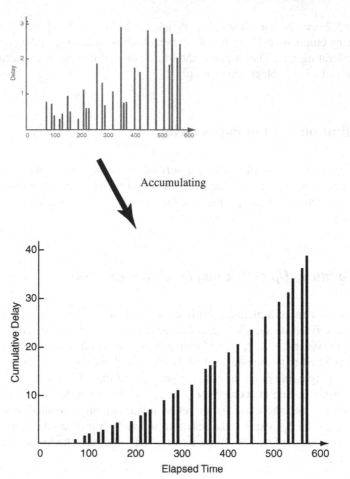

Fig. 5.2 Observation of cumulative delay

be defined by the amount $\delta_{jk}(S_i^A)$ given by

$$\delta_{jk}(S_i^A) = \max\left[C_{jk}(S_i^A) - C_{jk}(S_i^P), 0\right], \tag{5.3}$$

where $C_{jk}(*)$ signifies the completion time of task ϕ_{jk} on the schedule $*$. If $\delta_{jk}(S_i^A) > 0$, then task ϕ_{jk} is called a *delayed task* on $(\tau_{i-1}, \tau_i]$. It should be noted in the above that Eq. (5.3) counts up delays only for the tasks which complete during $(\tau_{i-1}, \tau_i]$.

A *total delay* $\hat{D}_i$ over $(\tau_{i-1}, \tau_i]$ is then given by

$$\hat{D}_i = \sum_{k=1}^{m} \sum_{\lambda=1}^{\Lambda} \delta_{j_\lambda k}(S_i^A), \tag{5.4}$$

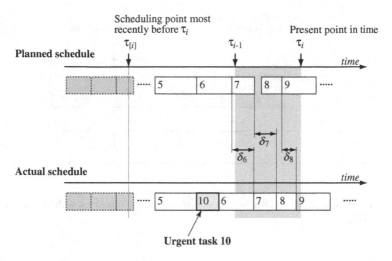

Fig. 5.3 Example of cumulative delay

and hence, a *cumulative delay* D_i at the inspection time τ_i becomes:

$$D_i = \sum_{j=l}^{i} \hat{D}_j, \tag{5.5}$$

where l is defined by $\tau_{[i]} = l\tau (l < i)$. When it is decided to conduct scheduling at the present time τ_i, we have $D_i = 0$.

Figure 5.3 depicts an example of a cumulative delay over the elapsed time. In Fig. 5.3, it is supposed that, at the current inspection point τ_i, processing of tasks 6, 7, and 8 were delayed due to the urgent task 10. Under such a situation, the total delay $\hat{D}_i$ over $(\tau_{i-1}, \tau_i]$ can be measured as $\delta_6 + \delta_7 + \delta_8$ because these three tasks 6, 7, and 8 completed during the time period $(\tau_{i-1}, \tau_i]$. If no unexpected events other than the urgent task 10 have occurred during the period $(\tau_{[i]}, \tau_i]$, the current cumulative delay D_i is equivalent to $\hat{D}_i$.

In the above, only the jobs or tasks completing during $(\tau_{i-1}, \tau_i]$ are detected and observed to count up their delays. From the managerial point of view, this approach is advantageous because of its simple managerial operations. Nevertheless, when some jobs or task should eminently be delayed beyond a long period of time, their delays would neither be measured nor be counted up for a long time. This is disadvantageous to the management since it may irretrievably be too late to take some countermeasures against such significant delays. In the succeeding subsection, another definition of a cumulative delay is introduced under which we can detect important delays more promptly.

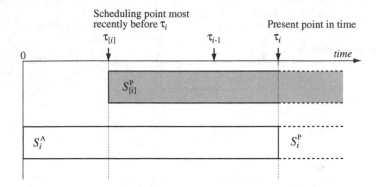

Fig. 5.4 Revised, actual, and predictive schedules observed at τ_i (reprinted with kind permission from ISCIE [2])

5.2.2 Counting Up at Every Inspection Point

In the previous subsection, the concept of a cumulative delay has been defined on the basis of individual task delays. It is, however, possible to define it in terms of job delays likewise, and this subsection develops another definition of a cumulative delay based on job delays [2].

Let us reuse the notation, $\tau_{[i]}$ ($\leq \tau_{i-1}$), which has been defined in Sect. 5.2.1. Additionally, we introduce the following notations:

J_i^A = a set of completed jobs at τ_i.

J_i^P = a set of uncompleted jobs at τ_i.

S_i^A = the actual schedule realized over the period $(0, \tau_i]$ which may include some delays.

S_i^P = the predictive schedule after τ_i resulting from the schedule revision at $\tau_{[i]}$.

Consider a job j on the schedule $S_{[i]}^P$ given by

$$j \in \tilde{J}_{[i]}^P = \{J_{[i]}^P | C_j(S_{[i]}^P) \leq \tau_i, \, C_j(S_i^A) > \tau_{i-1}\},$$

where $\tilde{J}_i^P$ represents a set consisting of jobs defined on $S_{[i]}^P$ which have not been finished at τ_{i-1} but are supposed to be completed before τ_i, while $C_j(*)$ denotes the completion time of job j on a schedule $*$. Figure 5.4 illustrates the predictive schedule $S_{[i]}^P$ obtained at $\tau_{[i]}$, the actual schedule S_i^A realized up to the present inspection point in time τ_i and the subsequent predictive schedule S_i^P starting at τ_i.

Let us express, by δ_j^i, the delay of job j in $\tilde{J}_i^P$ ($\neq \phi$) over the period $(\tau_{i-1}, \tau_i]$, and then observe δ_j^i, assuming that τ_i and $(\tau_{i-1}, \tau_i]$ are the current inspection point in time and the current period, respectively. The delay δ_j^i of job j can be discussed

Fig. 5.5 Case 1: Job j was to complete before τ_{i-1} on $S^P_{[i]}$; however, job j is still being processed at τ_i. Obviously, the delay δ^i_j of job j over the period $(\tau_{i-1}, \tau_i]$ agrees with the inspection time interval $\tau\ (=\tau_i - \tau_{i-1})$. The actual completion time of job j is not revealed at τ_i, thus it is anticipated that the inequality $\tau_i > C_j(S^A_i)$ holds

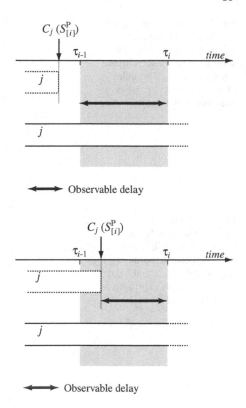

Fig. 5.6 Case 2: Job j was to be completed during $(\tau_{i-1}, \tau_i]$. However, job j is still being processed at the current point in time τ_i. the delay δ^i_j of job j of the period $(\tau_{i-1}, \tau_i]$ is given by $\tau_i - C_j(S^P_{[i]})$. The actual completion time of job j is not revealed at τ_i; thus, it is anticipated that the inequality $\tau_i > C_j(S^A_i)$ holds

according to the following four cases listed below:

(Case 1) The first case corresponds to a situation where job j is found not to have been completed yet at τ_i although it was scheduled to be finished before τ_{i-1} on the planned schedule $S^P_{[i]}$, i.e., $C_j(S^P_{[i]}) \leq \tau_{i-1}$. In this case job j is in process of being delayed, and the amount δ^i_j of its delay observed over the period $(\tau_{i-1}, \tau_i]$ becomes identical with the inspection time interval $\delta^i_j = \tau_i - \tau_{i-1} = \tau$ as shown in Fig. 5.5.

(Case 2) The second case explains a situation where job j was scheduled to be finished within the period $(\tau_{i-1}, \tau_i]$, but it is found, at τ_i, to be still being processed. The delay δ^i_j of job j over the current period is observed to be $\tau_i - C_j(S^P_{[i]})$ as depicted in Fig. 5.6.

(Case 3) The third case represents a situation under which job j was scheduled to be completed before τ_{i-1}, but it has just been completed within the current period $(\tau_{i-1}, \tau_i]$. This situation reveals that the delay of job j detected over the current period is given by $C_j(S^A_i) - \tau_{i-1}$ as shown in Fig. 5.7.

(Case 4) The last case is a situation where both the planned completion time and the actual one of job j are within the period $(\tau_{i-1}, \tau_i]$. In this case, we have the actual delay of job j given by $\max(C_j(S^A_i) - C_j(S^P_{[i]}), 0)$ as indicated in Fig. 5.8.

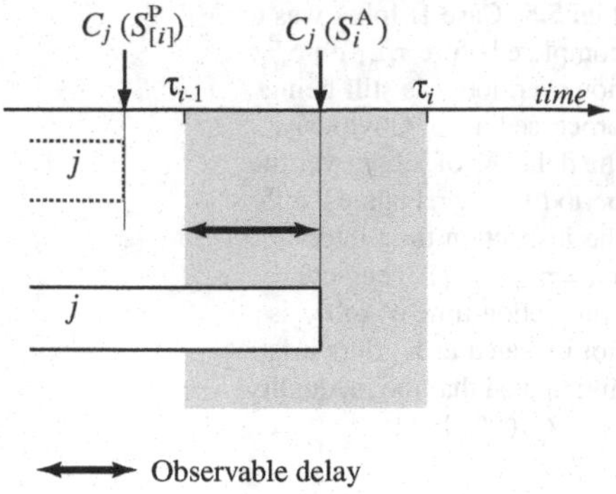

Fig. 5.7 Case 3: Job j was to be completed before τ_{i-1} on $S^P_{[i]}$, and then job j has been completed before the current point in time τ_i. The delay of job j over $(\tau_{i-1}, \tau_i]$ is observed to be $C_j(S^A_i) - \tau_{i-1}$

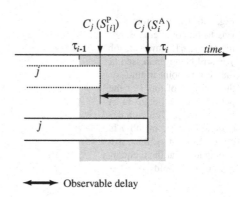

Fig. 5.8 Case 4: Job j was to be completed during $(\tau_{i-1}, \tau_i]$, and it has actually been completed during this period. The actual delay over $(\tau_{i-1}, \tau_i]$ is expressed as $C_j(S^A_i) - C_j(S^P_{[i]})$

In accordance with the above four cases, a *cumulative delay* D_i at τ_i can be defined by

$$D_i = D_{i-1} + \sum_{j \in \bar{J}^P_{[i]}} \delta^i_j \tag{5.6}$$

The delay δ^i_j observed in the above can be summarized as follows:

$$\delta^i_j = \begin{cases} 0, & \text{if } C_j(S^P_{[i]}) \leq C_j(S^A_i), \\ \tau \ (= \tau_i - \tau_{i-1}), & \text{if } C_j(S^P_{[i]}) \leq \tau_{i-1} \text{ and } C_j(S^A_i) > \tau_i, \\ \tau_i - C_j(S^P_{[i]}), & \text{if } C_j(S^P_{[i]}) > \tau_{i-1} \text{ and } C_j(S^A_i) > \tau_i, \\ C_j(S^A_i) - \tau_{i-1}, & \text{if } C_j(S^P_{[i]}) \leq \tau_{i-1} \text{ and } C_j(S^A_i) \leq \tau_i, \\ C_j(S^A_i) - C_j(S^P_{[i]}), & \text{if } C_j(S^P_{[i]}) > \tau_{i-1} \text{ and } C_j(S^A_i) \leq \tau_i. \end{cases} \tag{5.7}$$

It should be noted in the above that $C_j(S_i^A) > \tau_i$ signifies that job j has not been completed by τ_i.

Let us see a simple example explaining how to count up job delays based on Eq. (5.7) by utilizing Fig. 5.9. The upper chart in Fig. 5.9 illustrates how to measure the cumulative delay. In this figure, suppose that a focal job, say job j, was scheduled to be completed at a certain time within the period (τ_{20}, τ_{21}), but it has been delayed for some reason. In the following, notations C_j^P and C_j^A are introduced to, respectively, denote the planned completion time and the actual completion time of job j.

According to the definition given by Eq. (5.7), the delay of job j, at τ_{21}, was found to be

$$\delta_j^{21} = \tau_{21} - C_j^P.$$

At the next inspection point τ_{22}, it is revealed the delay of job j became at least $\delta_j^{21} + \tau$ because job j had not been completed yet at τ_{22}. Thus, the delay of job j over the period $(\tau_{21}, \tau_{22}]$ is given by

$$\delta_j^{22} = \tau,$$

where τ denotes the inspection time interval.

In the same way, the delay of job j is accumulated to yield

$$\delta_j^{23} = \delta_j^{24} = \delta_j^{25} = \delta_j^{26} \equiv \delta.$$

It has been revealed, at the inspection time τ_{27}, that job j was completed within the period $(\tau_{26}, \tau_{27}]$. The delay δ_j^{27} of job j over $(\tau_{26}, \tau_{27}]$ results in

$$\delta_j^{27} = C_j^A - \tau_{26},$$

using the actual completion time of job j.

On the other hand, the lower chart of Fig. 5.9 describes how the delays at individual inspection points from τ_{21} to τ_{27} are accumulated.

The actual amount of delay of job j in Eq. (5.1) is eventually obtained at the inspection point τ_{27} by summing up δ_j^i since

$$
\begin{aligned}
D(j) &= \sum_{l=21}^{27} \delta_j^l \\
&= \tau_{21} - C_j^P + 5\tau + C_j^A - \tau_{26} \\
&= C_j^A - C_j^P + 5\tau - (\tau_{26} - \tau_{21}) \\
&= C_j^A - C_j^P.
\end{aligned}
$$

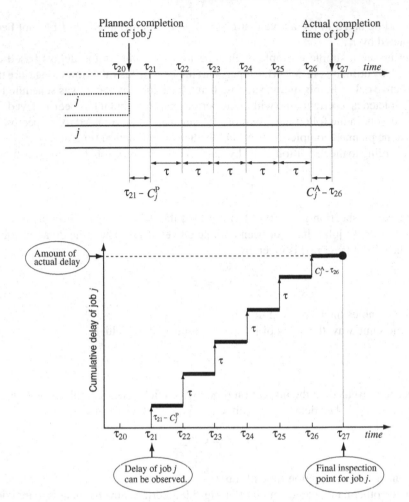

Fig. 5.9 Example of cumulative delay based on inspection interval (reprinted with kind permission from ISCIE [2])

5.3 Basic Properties of Cumulative Delay

This section demonstrates basic properties of a cumulative delay defined in Sect. 5.2.1 by performing computational simulations under typical job shop configurations with random machine breakdowns [4]. In Sect. 6.3, more elaborate simulations will be conducted for the same purpose and therefore the readers can refer to it for details. With a view to observing the behavior of delays, suppose that no schedule revision is conducted except the right-shift operation applied to the predictive schedule of each instance when machine breakdowns occur. The schemes of scheduling simulations are as follow:

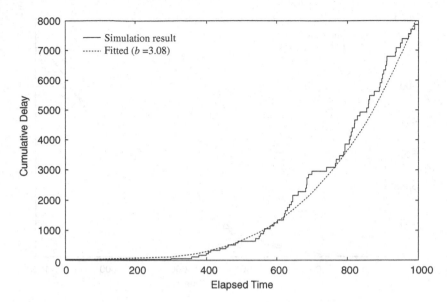

Fig. 5.10 Mean cumulative delay $[(m, n) = (10, 100)]$ (reprinted with kind permission from Elsevier [4])

Generation of Instances: An instance is randomly generated for each of the two types of job shop problems with $(m, n) = (10, 100)$ and $(m, n) = (5, 200)$, where m denotes the number of processing machines while n represents the number of jobs. In each of the two instances, all the jobs are assumed to be ready to enter the manufacturing system at time zero, and the number of tasks of which each job consist is obtained by generating random integers following uniform distribution Uniform($\lceil m/2 \rceil, m$), where $\lceil * \rceil$ is the smallest integer greater than $*$. The processing time of each task is, likewise, obtained by generating random integers which follow a uniform distribution, Uniform($0, 20$).

Machine breakdown scenarios: The number of machines subject to breakdowns is given by $\lceil \alpha m \rceil$, and $\alpha = 0.7$, e.g., signifies 70 % of machines are possibly down during execution of the schedule. It is assumed that the times between breakdowns can be represented in terms of independent and identically distributed (*iid*) random variables having the exponential distribution EXP($\beta \overline{p_m}$) with mean $\beta \overline{p_m}$, where $\beta \overline{p_m}$ denotes the expected total processing time on each machine. Further, repair times are *iid* random variables having the log-normal distribution LN($\ln(\gamma_1 \overline{p_{jk}}), \gamma_2$), where $\ln(\gamma_1 \overline{p_{jk}})$ and γ_2, respectively, denote its mean and variance and $\overline{p_{jk}}$ is the mean processing time of a task.

For each instance, considered are two types of machine breakdown behaviors:

- The instance of a problem $(10, 100)$: $(\alpha, \beta, \gamma_1, \gamma_2) = (0.7, 0.1, 1.0, 0.01)$
- The instance of a problem $(5, 200)$: $(\alpha, \beta, \gamma_1, \gamma_2) = (0.7, 0.1, 1.0, 0.05)$

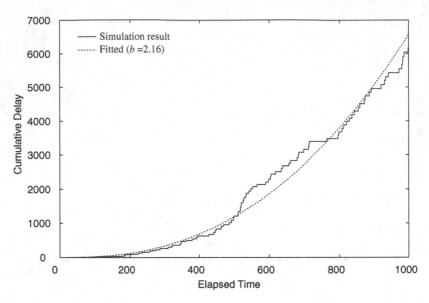

Fig. 5.11 Mean cumulative delay $[(m, n) = (5, 200)]$ (reprinted with kind permission from Elsevier [4])

Computational simulations of machine breakdowns were carried out 50 times for each of the two instances.

The collected data from the simulation results are the cumulative delays at the actual completion time of each individual task. In addition, the average of cumulative task delays was also computed over 50 times repetitions when any task was completed in each individual problem instance. This is called the *mean cumulative task delay* in the following.

Figures 5.10 and 5.11 show the behavior of the mean cumulative task delays in the two instance problems (10, 100) and (5, 200), respectively. The solid lines in these figures represent the behaviors of the mean cumulative task delay caused by machine breakdowns against the elapsed time. Obviously, the behaviors of the mean cumulative task delay in these two figures show similar behavior to each other.

Next, a smooth curve was fitted to the line of mean cumulative task delays. The curve adopted here is given by

$$D(t) = at^b \quad (a > 0, b \geq 1), \tag{5.8}$$

where t denotes elapsed time, and both a and b are parameters. Particularly, parameter b determines the shape of the curve. The curve in Eq. (5.8) was adopted since it has a very simple structure. Parameters a and b involved in Eq. (5.8) were estimated from the collected data in relation to the mean cumulative task delays by means of the

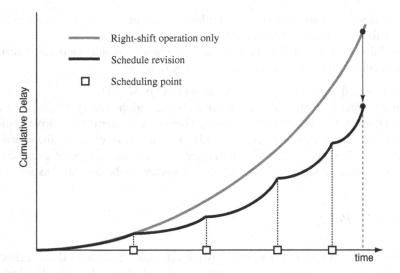

Fig. 5.12 Effectiveness of online scheduling from the viewpoint of cumulative delay

least square method. The resulting estimates, $\hat{b}$, of parameter b are

- $\hat{b} = 3.08$ for the instance of a problem (10, 100),
- $\hat{b} = 2.16$ for the instance of a problem (5, 200),

which indicate that both the estimates exceed 2, accordingly they satisfy the condition of $b \geq 1$. The resulting curve is represented as a dotted line in both Figs. 5.10 and 5.11.

From the above results, it is observed that the cumulative task delay tends to increase with elapsed time at an accelerating rate. This tendency implies that tasks undergoing delays affect most of their succeeding tasks and will cause a serious delay to the whole schedule as a result. Moreover, the nonlinear characteristics of the mean cumulative task delay, and hence, the cumulative task delay suggests the effectiveness of schedule revisions, i.e., schedule revisions, if they are executed appropriately, could effectively reduce the costs associated with schedule delays. Figure 5.12 depicts the effectiveness of schedule revisions based on the cumulative delay.

5.4 Configuration of Scheduling Policy

By employing the idea of the cumulative task delay or the cumulative job delay described above, a new approach to the problem when-to-revise against uncertainty would become available. Its underlying concept is as follows:

- Implement the monitoring of the cumulative delay on the schedule, for example, in the form of inspection points in time;
- Schedule revisions are to be performed if the current cumulative delay exceeds a prespecified threshold called *critical cumulative delay*.

This is identical with the concept of a *control limit policy* [1].

Let D^* denote the threshold mentioned above, and the policy proposed here is called D^*-*driven* (*schedule revision*) *policy* hereafter. Under the D^*-driven policy, an event that the cumulative delay exceeds D^* is a sort of event for the schedule; accordingly D^*-driven policy can be regarded as a specific type of event-driven policy described in Sect. 3.4.2. The D^*-driven policy can be described as follows:

D*-Driven Policy

(1) Inspections are executed to the existing schedule in order to detect schedule delays at inspection times $\tau_i = i\tau (\tau > 0, \ i = 1, 2, \ldots, M)$ over the planning horizon $(0, H]$ with $M\tau \leq H$.
(2) When the cumulative delay at τ_i is found to exceed D^*, a schedule revision is performed to the current schedule, and τ_i becomes the latest schedule revision point in time and D_i becomes equal to zero. This operation is identical with renewing operation of the schedule.

The main advantage of D^*-driven policy is that the decision-making process to conduct schedule revisions is simplified compared with typical event-driven scheduling policy since we can confine our attention to monitor only an increase of a cumulative delay.

Section 3.5 extended the two basic schedule revision policies, the periodic policy and the event-driven one to another two advanced policies named the hybrid policy [3, 5] and the enhanced event-driven one. The D^*-driven policy proposed in the above can be built in the two advanced policies to yield the *hybrid D^*-driven policy* and the *enhanced D^*-driven policy*. Their frameworks are as follows:

Hybrid D*-Driven Policy

(1) Inspections are executed to the existing schedule in order to detect schedule delays at inspection times $\tau_i = i\tau (\tau > 0, \ i = 1, 2, \ldots, M)$ over period $(0, H]$ with $M\tau \leq H$.
(2) A decision making in reference to whether or not to revise the current schedule is conducted at $T(= l\tau), 2T, \ldots,$ or LT where $LT \leq H, (L+1)T > H$. If it is determined to execute a schedule revision, some suitable method for revising the current schedule is adopted to do so, and τ_i becomes the latest schedule revision point.

(3) When the cumulative delay at τ_i is found to exceed a prespecified threshold called a critical cumulative delay D^*, a schedule revision is also carried into execution at inspection point τ_i, which becomes the latest schedule revision point and D_i is set to zero.

Enhanced D^*-Driven Policy

(1) Inspections are executed to the existing schedule for the purpose of detecting some schedule delays at inspection times $\tau_i = i\tau(\tau > 0,\ i = 1, 2, \ldots, M)$ over period $(0, H]$ with $M\tau \leq H$.
(2) When the elapsed time reaches T at τ_i since the previous schedule revision, then a schedule revision is enforced at the time, which becomes the latest schedule revision point.
(3) Even if the elapsed time does not reach T since the previous schedule revision, a schedule revision is executed to the current schedule when the cumulative delay at τ_i is found to exceed a prespecified threshold D^*. In this case, τ_i becomes the latest schedule revision point and D_i is set to zero.

References

1. Derman C (1966) Denumerable state Markovian decision processes—average cost criterion. Ann Math Stat 37:1545–1553
2. Suwa H, Fujimura D, Sandoh H (2004) Cumulative delay based reactive scheduling policy—its application to single-machine dynamic scheduling. Trans ISCIE 17(9):371–378
3. Suwa H (2007) A new when-to-schedule policy in online scheduling based on cumulative task delays. Int J Prod Econ 110:175–186
4. Suwa H, Sandoh H (2007) Capability of cumulative delay based reactive scheduling for job shops with machine breakdowns. Comput Ind Eng 53:63–78
5. Vieira GE, Herrmann JW, Lin E (2000) Analytical models to predict the performance of a single-machine system under periodic and event- driven rescheduling strategies. Int J Prod Res 38(8):1899–1915

Part III
Online Scheduling Based
on Cumulative Delay

Chapter 6
D^*-Driven Policy: Application to Job Shop Problems

Abstract This chapter introduces the D^*-driven policy which was briefly described in the previous chapter, considering job shop environments with random machine breakdowns as disruptions. Disjunctive representations of job shop problems with random breakdowns are first discussed, and then the properties of the cumulative task delay are elaborately investigated through computational simulations. The applicability and the capability of the D^*-driven policy are also discussed.

6.1 Job Shop Scheduling with Disruptions

We here consider a method for modeling a job shop environment with random machine breakdowns. Let us confine ourselves to a problem in which a set of jobs are to be processed on a set of machines. The number of jobs is denoted by n, while m signifies the number of machines. In addition, job j ($j = 1, 2, \ldots, n$) has its own due date, d_j.

It is assumed that each machine can process only one job at one time, and preemption of job processing is not allowed except when machine breakdowns occur. Each individual job consists of n_j tasks, $\phi_{j\mu(1)}, \phi_{j\mu(2)}, \ldots, \phi_{j\mu(n_j)}$, where $\mu(l)$ ($l = 1, 2, \ldots, n_j$, $\mu(l) \in \{1, \ldots, m\}$) expresses a machine on which the lth task of job j is to be processed. In the above, we have introduced $\mu(l)$ to specify the machine on which the lth task is to be processed. More precisely, $\mu(l)$ should be written as $\mu_j(l)$ so that the machine to be assigned to the task can uniquely be identified. The notation, $\mu(l)$, however, always follows j for job identification, accordingly we employ the notation, $\mu(l)$, hereafter to avoid notational nuisances and intricacies.

Every two tasks of job j have a precedence relation between each other, that is, task $\phi_{j\mu(l'+1)}$ ($l' = 1, 2, \ldots, n_j - 1$) cannot start until task $\phi_{j\mu(l')}$ is completed. The job routing, i.e., the technological sequence of the tasks in each job, the due date, d_j, of job j, and the processing time, $p_{j\mu(l)}$, of task $\phi_{j\mu(l)}$ are known in advance. Let $e_{j\mu(l)}(s)$ denote the starting time of $\phi_{j\mu(l)}$ on schedule s.

In this chapter, we consider the maximum tardiness, L_{max}, as the primary performance measure of the schedule. Let S_0 and C_{max} respectively denote a predictive schedule starting at time zero and its makespan for a given job shop problem. When machine breakdowns occur to make S_0 infeasible during its process, S_0 will naturally be modified by means of some suitable method for scheduling.

6.2 Disjunctive Representation

6.2.1 Basic Model

Let us represent a general job shop scheduling problem using a *disjunctive graph* $G = (V, C, D)$, where V expresses a node set, and C and D respectively denote a set of conjunctive (directed) arcs and that of disjunctive arcs. The node set, V, consists of the following four kinds of nodes:

- A *task node* $j(l)$, which corresponds to the lth task, $\phi_{j\mu(l)}$, of job j to be processed on machine $\mu(l) \in \{1, \ldots, m\}$.
- A *task starting node* $\tilde{j}$ that represents the starting state of the first task of job j.
- A *job completion node* $\underline{j}$ that signifies the completed state of job j, i.e., that of the last task of job j.
- The source and sink nodes denoted by 0 and by $*$, respectively.

In addition, let (u, v) $(u, v \in V, u \neq v)$ denote a conjunctive (directed) arc emanating from node u to node v. The source node, 0, has several conjunctive arcs with length zero emanating from itself, each of which enter one of the task starting nodes, $\tilde{j}$, of job j. The task starting node, $\tilde{j}$, of job j that corresponds to the first task in job j, also has a conjunctive arc $(\tilde{j}, j(l''))$ of length zero emanating from itself into task node $j(l'')$, where $j(l'')$ represents task $\phi_{j\mu(l'')}$. The sink node, $*$, has several conjunctive arcs with length $K - d_j$ emanating from a job completion node $\underline{j}$ into itself, where $K \geq \max_j\{d_j\}$ [1].

The task completion node, $\underline{j}$, of job j has a conjunctive arc emanating into itself from $j(n_j)$, where $j(n_j)$ represents a task $\phi_{j\mu(n_j)}$, the last task in job j.

The arc set, C, consists of not only the above four types of conjunctive arcs, but conjunctive arcs, $(j(l'), j(l' + 1))$ $(l' = 1, 2, \ldots, n_j - 1)$, of job j, where $(j(l'), j(l' + 1))$ expresses a precedence relation between the two tasks, $\phi_{j\mu(l')}$ and $\phi_{j\mu(l'+1)}$, coming from a technological order for processing. All the conjunctive arcs emanating from node $j(l)$ $(l = 1, 2, \ldots, n_j)$ have a length equal to the processing time, $p_{j\mu(l)}$, of the corresponding task, $\phi_{j\mu(l)}$.

On the other hand, the disjunctive set, $D = \{\{u, v\}|u, v \in V\}$, consists of pair of conjunctive arcs $\{u, v\} = \{(u, v), (v, u)\}$. In the case of $\{j_1(l_1), j_2(l_2)\} \in D$, the two tasks, $\phi_{j_1\mu(l_1)}$ and $\phi_{j_2\mu(l_2)}$, which are respectively expressed by nodes $j_1(l_1)$ and

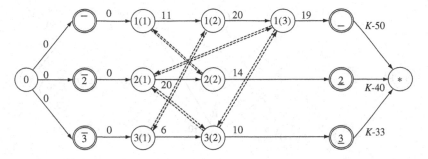

Fig. 6.1 Disjunctive graph for instance P1 (reprinted with kind permission from ISCIE [4])

Table 6.1 Instance P1 with $m = 3$ and $n = 3$ (reprinted by permission from ISCIE [4])	Job j	n_j	$\phi_{jk}(p_{jk})$	Due date d_j
	1	3	ϕ_{12} (11), ϕ_{11} (20), ϕ_{13} (19)	50
	2	2	ϕ_{23} (20), ϕ_{22} (14)	40
	3	2	ϕ_{31} (6), ϕ_{33} (10)	33

$j_2(l_2)$, are to be processed on the same machine. In other words, $\mu(l_1)$ of job j_1 equals $\mu(l_2)$ of job j_2. To select and adopt either one of the two conjunctive arcs, (u, v) and (v, u), is equivalent to determine the processing order between the two tasks, u and v, on the same machine.

Consequently, choosing of either one of the two conjunctive arcs from individual elements of D without yielding a cyclic directed graph coincides with obtaining a feasible schedule. An acyclic directed graph which corresponds to a feasible schedule s, is referred to as a feasible graph denoted by G_s. In the feasible graph, G_s, the length of the longest path from the source node, 0, to the sink node, $*$, is called the critical path. It follows that the feasible graph, G_{s*}, which has the minimum value of $L_{\max}(s^*)$, provides an optimal schedule s^*.

Figure 6.1 depicts an example of a disjunctive graph for the instance P1 summarized in Table 6.1 [4]. In Fig. 6.1, the value of the arc length is displayed beside its corresponding arc. A feasible graph obtained from Fig. 6.1 is shown in Fig. 6.2 along with its corresponding feasible schedule. The maximum tardiness of this schedule is found to be zero.

6.2.2 Disjunctive Model with Machine Breakdown

Suppose that a schedule S over a period $(0, H]$ was initially given at time zero with a feasible graph $G(S)$ and that we have already processed a part of S. In this subsection, we first overview a method for regenerating a feasible schedule S' for the remaining part of S.

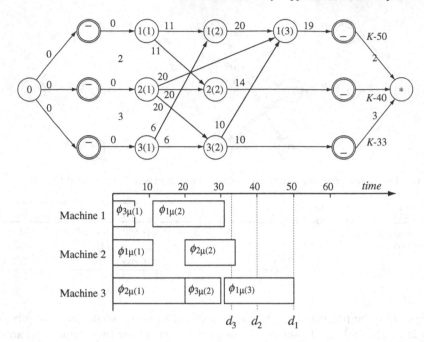

Fig. 6.2 Example of a directed graph with its resulting schedule for instance P1 (reprinted with kind permission from ISCIE [4])

It is assumed that a new schedule after τ_i is required for the remaining part of S. Let $\phi_{j\mu(l*)}$ be the task of job j ($j = 1, 2, \ldots, n$) having the earliest starting time after τ_i, and then, by concentrating on the tasks, $\phi_{j\mu(l*)}$, the procedure to obtain S' from S can be summarized as follows:

(1) Eliminate, from $G(S)$, the task nodes corresponding to the completed tasks up to τ_i, all the directed arcs emanating from these task nodes, and all the directed arcs emanating from their relevant task starting nodes.

(2) Append a directed arc $(0, j(l^*))$ with the length, $e_{j\mu(l*)}(S')$, in order to connect each starting node, $\tilde{j}$, to its associated task node corresponding to $\phi_{j\mu(l*)}$ for individual jobs.

In the following, the above procedure is called the *reconfiguration procedure*.

Figure 6.3 illustrates the initial schedule, S, along with its relevant graph, $G(S)$, and the resulting graph, $G(S')$, from the above procedure. In this figure, the four tasks, $\phi_{3\mu(1)}$, $\phi_{1\mu(2)}$, $\phi_{1\mu(1)}$, and $\phi_{2\mu(1)}$, have been eliminated from $G(S)$ to yield the new feasible graph, $G(S')$, since they have already been completed or are still in progress at τ_i. In $G(S')$, each arc emanating from a starting node into its relevant task node has a positive length.

Let us consider a right-shift operation as well as the computation of task delays at a point in time τ_i. It is assumed that a breakdown occurred on machine k ($k \in (1, 2, \ldots, m)$) at time t_b as a disruption to cause delays, and that machine k recovered

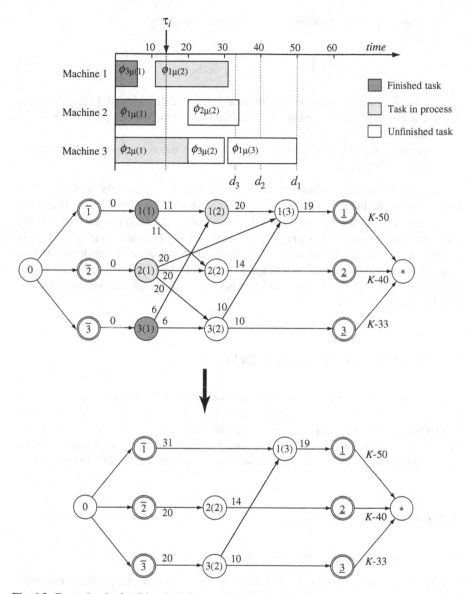

Fig. 6.3 Example of a feasible schedule at τ_i for instance P1 without uncertainty (reprinted with kind permission from ISCIE [4])

at t_r $(t_b < t_r \leq \tau_i)$. Then we can obtain a new feasible schedule after τ_i in the following manner:

1. Select a task, say $\phi_{j^*\mu(l)}$, of job j^* which has the earliest starting time among tasks to be processed on machine k even if partially over period $[t_b, t_r]$. Calculate

an amount of delay Δ of task $\phi_{j^*\mu(l)}$, which is given by

$$\Delta = \begin{cases} t_r - t_b \, (\equiv \lambda), & \text{if } e_{j^*\mu(l)}(S') \leq t_b \\ t_r - e_{j^*\mu(l)}(S'), & \text{otherwise} \end{cases}. \tag{6.1}$$

2. For the directed arc emanating from node $j^*(l)$, update its length to $(p_{j^*\mu(l)} + \Delta)$ on the existing feasible graph, and then recalculate the longest path of the newly generated, $G_{S'}$.
3. Generate a feasible graph after τ_i according to the reconfiguration procedure mentioned before.

Figure 6.4 illustrates a Gantt Chart in which task $\phi_{1(1)}$ was delayed four unit of time owing to the breakdown on machine 2 with a down time period $[0, 4]$. The graph below the Gantt chart expresses the schedule after a right-shift operation following the above procedure. In this graph, each length of the two arcs from node 1(1) is shown in a square to reveal it has been updated in response to the delay. At this stage, the schedule reflected by this graph is not feasible. Figure 6.4 also shows, in its bottom, the final graph generated by applying the reconfiguration procedure to yield a feasible schedule.

6.3 Properties of Cumulative Delay

The definition of a cumulative delay and its computational procedure were presented in Sect. 5.2. The basic properties of the cumulative delay were also overviewed through the results of computational simulations in Sect. 5.3. This section discusses its properties more deeply than Sect. 5.3 through computational experiments in the job shop environments, and explores a method for designing the D^*-driven policy.

6.3.1 Simulation Schemes

The simulation schemes to be used in this section are as follows [5]:

Problem instances: Table 6.2 summarizes a scheme to generate problem instances. We consider two values $m = 4$ and 8 for the number of machines, and three values $n = 60, 90$, and 120 for the number of jobs. It is assumed that all the jobs are ready to be processed at time zero.

The number, $n_j (j = 1, 2, \ldots, n)$, of tasks of job j is determined by generating a random integer following the discrete uniform distribution over an interval $[\lceil m/2 \rceil, m]$, where $\lceil * \rceil$ is the smallest integer greater than $*$. In the following, Uniform$(*, \cdot)$ represents the discrete uniform distribution over an interval $[*, \cdot]$. The processing time, p_{jk}, of task ϕ_{jk} is, likewise, obtained by generating a random integer following Uniform$(5, 10)$.

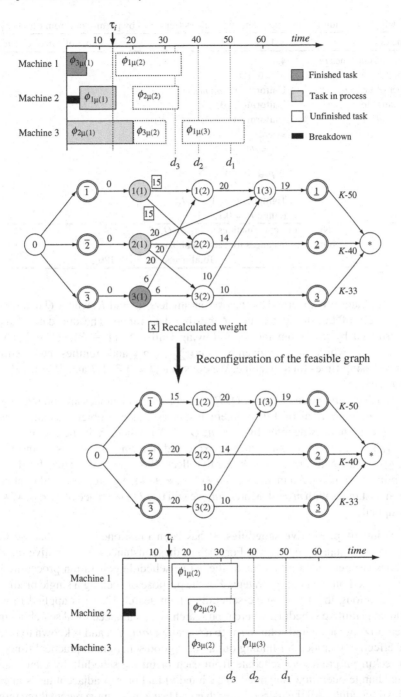

Fig. 6.4 Generation of right-shift schedule at τ_i for instance P1 (reprinted with kind permission from ISCIE [4])

Table 6.2 Generation of job shop problem instances (reprinted by permission from Elsevier [5])

Parameters	Values	Number of values considered
Number of machines m	4, 8	2
Number of jobs n	60, 90, 120	3
Number of tasks n_j	Uniform($\lceil m/2 \rceil, m$)	
Processing time p_{jk}	Uniform(5, 10)	
Due date d_j	Uniform ($\epsilon(1 - R), \epsilon(1 + R)$), where	
	$\epsilon = Q \cdot p_{\max},$	
	$p_{\max} = \displaystyle\max_{1 \leq k \leq m} \sum_{j=1}^{n} p_{jk}$	
	Tightness $Q = 1.3, 1.7,$	2
	Range $R = 0.1, 0.3$	2
	Number of combinations of m, n, Q, R	24
	Number of instances for each combination	5
	Total instances	120

Job due dates are governed by the two parameters, Q and R, where Q determines the tightness of the due dates and R the due date range. The due date of a job is determined by a random integer following Uniform($\epsilon(1 - R), \epsilon(1 + R)$) with $\epsilon = Q \cdot p_{\max}$, where $p_{\max}$ is given by $\max_k \sum_{j=1}^{n} p_{jk}$ and signifies the maximum total processing times on a machine. We consider $Q = 1.3, 1.7$ and $R = 0.1, 0.3$ in this subsection.

For each combination of (m, n, Q, R), five problem instances are randomly generated, yielding a total of 120 problem instances. Each problem instance generated in this manner is denoted by $I(m, n, Q, R, X)$, where X is the instance index ($X = 1, 2, \ldots, 5$). An instance class is expressed in terms of a symbol $*$ instead of m, n, Q, R, and X in $I(m, n, Q, R, X)$ to reflect all the possible values for the corresponding parameter. An instance class $I(4, *, *, *, *)$, e.g., represents the set of all problem instances with $m = 4$. Moreover, we will use $I(*)$ in place of $I(*, *, *, *, *)$ for simplicity.

Generation of predictive schedules: It has been mentioned above that we have 120 problem instances as a result. For each of these instances, 10 primitive feasible schedules are generated by means of the active schedule generation procedure [3], which is based on the greedy strategy. For the purpose of seeking a single predictive schedule among the 10 primitive schedules, tabu search [2, 3] is applied to each individual primitive schedule, where tabu search is an enhanced local search method by memorizing the visited solutions into a so-called *tabu list*, and is known to be one of the effective methods for finding out a fairly good schedule in practical time.

In seeking a predictive schedule from each primitive schedule by tabu search, an interchange operation is applied to each individual pair of adjacent tasks in each primitive schedule. All the pairs of interchanged tasks in the most recent three moves are used as attributes in the tabu list on the condition that the number of iterations is restricted to be at most 1,000 as a stop criterion. Among the 10 fairly good predictive

Table 6.3 Scheme for machine breakdown (reprinted by permission from Elsevier [5])

Parameters	Values	Number of values considered
Number of machines subjected to breakdown	αm, where	
	$\alpha = 0.5, 1.0$	2
Inter-breakdown times	$\tau_b \sim \text{EXP}[1/(\beta p_{max})]$, where	
	$\beta = 0.1, 0.3$	2
Down time	$\lambda \sim \text{LN}(\ln(\gamma_1 \overline{p_{jk}}), \gamma_2)$, where	
	$\gamma_1 = 0.5, 1.0,$	2
	$\gamma_2 = 0.01, 0.05$	2
	Number of scenarios	16
Total number of instance-scenario combinations		1920

schedules obtained in this manner, the best schedule is regarded as the single predictive schedule, S_0, for each instance problem. In addition, the makespan, C_{max}, and the maximum tardiness, L_{max}, are also computed in each predictive schedule since they are used in the performance evaluation of the D^*-driven policy.

Machine breakdowns: Table 6.3 describes the scenarios used for generating machine breakdown events. The number of machines subjected to breakdowns is given by $\lceil \alpha m \rceil$, where α is set to 0.5 or 1.0. When we have $\alpha = 0.5$, e.g., in the four machine environment, two of them are subjected to breakdowns. It is assumed that the breakdown times are independent and identically distributed (*iid*) random variables having the exponential distribution with mean βp_{max}, where β takes on the value 0.1 or 0.3. In the following, $\text{EXP}(*)$ represents the exponential distribution with a rate $*$ or with a mean $1/*$. Further, repair times are also *iid* random variables having the log-normal distribution with mean $\ln(\gamma_1 p_{..})$ and variance γ_2, which is denoted by $\text{LN}(\ln(\gamma_1 p_{..}), \gamma_2)$, where $p_{..}$ is the mean processing time of all the tasks. The parameter, γ_1, takes on the value 0.5 or 1.0, while γ_2 is set to 0.02 or 0.05.

These observations reveal that we have 16 breakdown scenarios. It follows that we have a total of $120 \times 16 = 1920$ problem instance-breakdown scenario combinations since we have the 120 problem instances. Let $B(\alpha, \beta, \gamma_1, \gamma_2)$ represent a breakdown scenario, where parameters α, β, γ_1, and γ_2 were defined in the above. The symbol $*$ is used again to indicate a breakdown class. A breakdown class $B(*, 0.3, *, *)$, e.g., represents the set of all breakdown scenarios with $\beta = 0.3$. We also use $B(*)$, for simplicity, instead of $B(*, *, *, *)$ to express the set of all the breakdown scenarios.

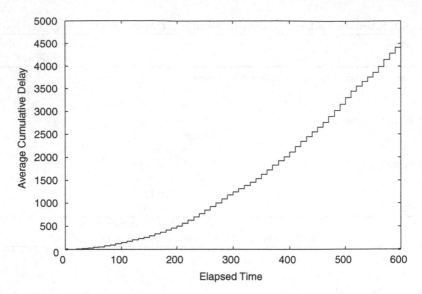

Fig. 6.5 Behavior of mean cumulative delay: the job shop instance with eight machines and 60 jobs

6.3.2 Simulation Results and Discussion

We conducted a simulation for each one of the individual 120 problem instances by applying each of the 16 breakdown scenarios to collect the data associated with cumulative delay. The average of cumulative task delay was also computed over the 16 machine breakdown scenarios at the time when any task was completed in each individual problem instance. The resulting average of the cumulative task delay is called the *mean cumulative (task) delay* in the following.

Figures 6.5 and 6.6 show two representative results of the 120 mean cumulative task delays. It is revealed in both figures that the mean cumulative task delay tends to increase at an accelerating rate with respect to the elapsed time. This is rhetorically the same as what we have mentioned in Sect. 5.3, but the findings here are described on the basis of more abundant and elaborate simulations in reference to the number of problem instances and machine breakdown scenarios. Besides, most of the properties of the mean cumulative task delay to be discussed are considered to hold true for the individual cumulative task delays as well. Their average is introduced simply because their behavior will be expressed in a relatively smooth curve with many and small steps.

In the same way as observed in Sect. 5.3, a smooth curve was fitted to each of the mean cumulative delays, two of which are shown in Figs. 6.5 and 6.6. The curve fitted to the resulting mean cumulative delay is given by the same equation as that in Sect. 5.3, that is, it is given by

$$D(t) = at^b \quad (a > 0, b \geq 1), \tag{6.2}$$

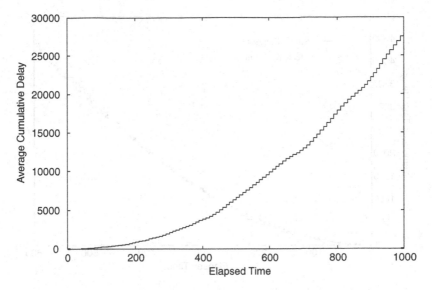

Fig. 6.6 Mean cumulative delay in the case of eight machines and 120 jobs

where t represents the elapsed time, and both a and b are parameters. In particular, note that b determines the shape of the curve. The least square method was used again to estimate the parameters a and b involved in Eq. (6.2) based on the data in relation to the cumulative delay.

Figure 6.7 shows the fitted curve, $D(t)$, as well as the behavior of the mean cumulative task delay in the case of 60 jobs in Fig. 6.5. It is seen in this figure the fitted curve can explain the behavior of the cumulative curve adequately, and this result motivated us to employ $D(t)$ in Eq. (6.2) as an analytical model of the cumulative delay. Table 6.4 shows the resulting estimates of parameter a for each instance, while those for parameter b are shown in Table 6.5. It is well known that the estimate of b significantly affects the shape of the curve and the estimate of a, accordingly we focus on the estimates of b in Table 6.5. Table 6.5 indicates that all the estimates of b take on the value close to 2.0, which imply that all the mean cumulative task delays obtained in this experiment sharply increases with elapsed time.

Let us examine the effects of parameters involved in the problem instances and machine breakdown scenarios upon the behavior of the mean cumulative task delay. For this purpose, computational experiments were further performed for the combinations of the 120 instances and the 16 breakdown scenarios. More precisely, we obtained a single predictive schedule in each of the 120 instances, applied each machine breakdown scenario to the individual predictive schedules, and then iterated simulation runs for each individual instance-scenario combination 50 times to collect data concerning cumulative task delays. The data were collected for each individual predictive schedule at the time when any task was completed in the pre-

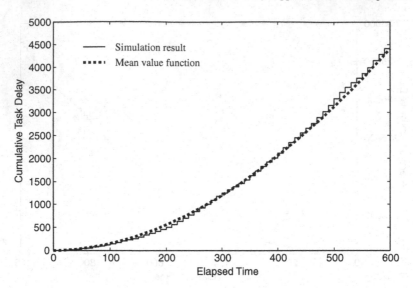

Fig. 6.7 Analytical model of cumulative delay

Table 6.4 Estimates of parameter a ($\times 10^{-2}$) in $D(t)$ for each instance (reprinted by permission from Elsevier [5])

Instance	$(Q, R) = (1.3, 0.1)$					$(Q, R) = (1.3, 0.3)$				
	1	2	3	4	5	1	2	3	4	5
(m, n)										
$(4, 60)$	3.57	6.22	2.93	2.99	2.56	2.74	0.65	3.11	5.25	2.16
$(4, 90)$	2.16	2.29	1.03	2.45	1.89	1.53	2.57	0.93	5.23	0.58
$(4, 120)$	2.45	1.13	3.56	1.10	0.85	2.17	1.72	1.28	0.91	0.72
$(8, 60)$	6.18	3.33	6.32	3.42	3.93	3.84	3.97	3.03	3.74	3.24
$(8, 90)$	1.06	1.29	1.90	3.34	5.29	4.72	3.39	1.27	3.51	1.47
$(8, 120)$	2.33	1.82	1.24	1.88	1.81	2.64	2.25	1.19	1.72	3.49
Instance	$(Q, R) = (1.7, 0.1)$					$(Q, R) = (1.7, 0.3)$				
	1	2	3	4	5	1	2	3	4	5
(m, n)										
$(4, 60)$	2.71	2.36	1.37	1.86	2.34	1.03	1.80	2.62	1.97	2.63
$(4, 90)$	2.71	2.81	1.90	2.02	0.43	0.62	0.69	2.89	2.24	1.40
$(4, 120)$	2.23	1.02	0.78	0.76	0.57	1.67	4.89	2.82	2.71	0.84
$(8, 60)$	6.06	3.02	4.14	3.09	4.90	5.82	4.52	6.84	1.81	3.21
$(8, 90)$	2.19	1.69	1.40	2.47	3.05	1.29	2.72	3.07	4.03	2.62
$(8, 120)$	1.30	3.32	3.85	2.99	0.92	1.53	1.48	1.47	5.49	1.52

Table 6.5 Estimates of parameter b in $D(t)$ for each instance (reprinted by permission from Elsevier [5])

Instances	$(Q, R) = (1.3, 0.1)$					$(Q, R) = (1.3, 0.3)$				
	1	2	3	4	5	1	2	3	4	5
(m, n)										
$(4, 60)$	1.822	1.752	1.881	1.909	1.907	1.929	2.158	1.882	1.793	1.940
$(4, 90)$	1.906	1.914	2.023	1.890	1.917	1.987	1.854	2.087	1.778	2.058
$(4, 120)$	1.830	1.961	1.762	1.983	1.976	1.843	1.891	1.933	2.000	2.031
$(8, 60)$	1.866	1.943	1.883	2.010	1.967	1.951	1.937	1.972	1.969	1.994
$(8, 90)$	2.076	2.106	2.004	1.927	1.864	1.863	1.899	2.079	1.918	2.064
$(8, 120)$	1.934	1.957	2.013	1.978	1.974	1.954	1.932	2.020	1.983	1.870
X	$(Q, R) = (1.7, 0.1)$					$(Q, R) = (1.7, 0.3)$				
	1	2	3	4	5	1	2	3	4	5
(m, n)										
$(4, 60)$	1.912	1.878	2.026	1.970	1.933	2.064	1.979	1.844	1.958	1.918
$(4, 90)$	1.855	1.833	1.959	1.880	2.145	2.084	2.063	1.850	1.892	1.961
$(4, 120)$	1.827	1.955	1.992	2.013	2.028	1.857	1.751	1.790	1.792	2.001
$(8, 60)$	1.877	1.963	1.966	1.971	1.880	1.895	1.925	1.854	2.055	1.988
$(8, 90)$	1.945	2.025	2.021	1.922	1.890	2.093	1.990	1.930	1.907	1.992
$(8, 120)$	2.007	1.864	1.833	1.866	2.031	2.001	2.019	2.031	1.777	2.007

dictive schedule. As a result, we obtained a single mean cumulative task delay per instance-scenario combination, in other words, we obtained a total of 1920 estimates of b.

Table 6.6 summarizes the relationship between the estimates of parameter b and the instance-scenario combinations. It is seen in Table 6.6 that the estimates of b reveal slightly larger values when the mean time between machine breakdowns is relatively smaller or when the mean repair time is relatively larger. Since $D(t)$ increases sharply when b takes on the larger value, these results indicate that the cumulative task delay tends to increase more rapidly if machine breakdowns occur more frequently or if their repairs take longer times. In this table, moreover, parameters γ_1 and γ_2 of a repair time distribution appear to have a larger impact on a cumulative task delay than α and β in relation to the machine breakdown frequency.

6.4 Performance of D^*-Driven Policy

6.4.1 Design of D^*-Driven Policy

Under the D^*-driven policy, a schedule revision is performed to the current schedule when the cumulative task delay exceeds a prespecified critical cumulative delay denoted by D^*. One of the design variables of the D^*-driven policy is evidently the critical cumulative delay, D^*. However, the inspection time interval, τ, can be

Table 6.6 Estimates of parameter b with respect to breakdown class (AVG: average, SD: standard deviation) (reprinted by permission from Elsevier [5])

Instance class—breakdown class	Estimates of b			
	AVG	SD	MIN	MAX
Number of machines (m)				
$I(4, *, *, *, *) - B(*)$	1.918	0.096	1.677	2.292
$I(8, *, *, *, *) - B(*)$	1.933	0.100	1.626	2.299
Number of jobs (n)				
$I(*, 60, *, *, *) - B(*)$	1.922	0.102	1.653	2.299
$I(*, 90, *, *, *) - B(*)$	1.934	0.098	1.626	2.287
$I(*, 120, *, *, *) - B(*)$	1.927	0.096	1.626	2.240
Due date tightness (Q)				
$I(*, *, 1.3, *, *) - B(*)$	1.937	0.094	1.653	2.287
$I(*, *, 1.7, *, *) - B(*)$	1.914	0.101	1.626	2.299
Due date range (R)				
$I(*, *, *, 0.1, *) - B(*)$	1.924	0.098	1.626	2.294
$I(*, *, *, 0.3, *) - B(*)$	1.927	0.099	1.653	2.299
Number of machines subjected to breakdowns (α)				
$I(*), B(0.5, *, *, *)$	1.947	0.103	1.653	2.299
$I(*), B(1.0, *, *, *)$	1.904	0.089	1.626	2.215
Mean time between breakdowns (β)				
$I(*), B(*, 0.1, *, *)$	1.933	0.086	1.715	2.242
$I(*), B(*, 0.3, *, *)$	1.918	0.109	1.626	2.299
Mean repair time (γ_1)				
$I(*), B(*, *, 0.5, *)$	1.877	0.081	1.626	2.287
$I(*), B(*, *, 1.0, *)$	1.974	0.090	1.698	2.299
Variance of repair times (γ_2)				
$I(*), B(*, *, *, 0.01)$	1.959	0.100	1.626	2.299
$I(*), B(*, *, *, 0.05)$	1.891	0.084	1.626	2.169

regarded as the other design variable unless it is exogenously determined by the monitoring system. From this point of view, we deal with both D^* and τ as design variables in the following. Since it is very difficult to acquire the optimal values for these two design variables analytically, we consider the following simulation-based approach simply for the purpose of investigating the properties of the D^*-driven policy by comparing it with the conventional event-driven policy later.

Using the results of the computational simulations we have observed, let us introduce a set of candidates for both of the two design variables in the following manner: Let $D_\eta^*(t)$ be given by

$$D_\eta^*(t) = [\hat{a} t^{\hat{b}}], \qquad (6.3)$$

where $\hat{a}$ and $\hat{b}$ are, respectively, the estimates of parameters a and b in instance $\eta(= I(m, n, Q, R, X))$ in Tables 6.4 and 6.5, which were obtained through the computational simulations in the previous subsection. By letting $t = 0.05 \times C_{\max}$, $0.1 \times C_{\max}$, $0.2 \times C_{\max}$, three unique values of $D_\eta^*(t)$ can be obtained in each instance problem, where $C_{\max}$ is the makespan of the predictive schedule in each instance. Let us denote by, $D_{1\eta}^*$, $D_{2\eta}^*$, and $D_{3\eta}^*$, these three unique values in each instance, i.e., we have

$$D_{1\eta}^* = D_\eta^*(0.05C_{\max}),$$
$$D_{2\eta}^* = D_\eta^*(0.1C_{\max}),$$
$$D_{3\eta}^* = D_\eta^*(0.2C_{\max}).$$

Let $\overline{D}_i^*$ ($i = 1, 2, 3$) denote the average of $D_{i\eta}^*$ computed over instances $\eta \in I(m, n, Q, R, *)$, and $\tilde{D}_i^*$ ($i = 1, 2, 3$) express the average of $D_{i\eta}^*$ over all the instances $\eta \in I(*)$.

On the other hand, let the inspection time interval, τ, be equal to:

$$\tau = 10, 20, 30 \text{ and } 60, \quad \text{e.g., counted by minutes,}$$

which are suitably determined by regarding a makespan $C_{\max}$ as 1,440, e.g., one day counted by minutes, in each instance.

In the above, we have generated three types of candidates, $D_{i\eta}^*$, $\overline{D}_i^*$, and $\tilde{D}_i^*$ ($i = 1, 2, 3$) for the critical cumulative delay, D^*, and $\tau = 10, 20, 30$ and 60 for the inspection time interval, τ. As a result, we have obtained a candidate set consisting of a total of 36 pairs of design variables D^* and τ. In the following, we examine the performance of the D^*-driven policy using the candidate set $\{(D^*, \tau)|D^* = D_{i\eta}^*, \overline{D}_i^*, \tilde{D}_i^*, i = 1, 2, 3, \tau = 10, 20, 30, 60\}$ of the design variables.

6.4.2 Simulation Schemes for Performance Evaluation

We here investigate the performance of the D^*-driven policy through computational simulations by comparing it with a conventional event-driven schedule revision policy (EDR in short). For this purpose, we apply these two types of policies to the job shop problems in Sect. 6.3.

As observed in Sect. 6.3, we have generated a predictive schedule in each of the 120 instances $I(m, n, Q, R, X)$, and then applied each one of the 16 machine break-down scenarios to the individual predictive schedules, yielding the 1920 instance-scenario combinations. To each one of the resulting combinations, we here apply the two types of schedule revision policies to concentrate on both the maximum tardiness, $L_{\max}$, and the number of schedule revisions carried out. Note that the EDR can be realized by the D^*-driven policy with $\tau = 1$, the minimum inspection interval, to reflect the situation where the existing schedule is monitored continuously in time. This is to draw forth the high performance of inhibiting the growth of the makespan or the maximum tardiness under the EDR.

Once a decision to conduct a schedule revision is made, e.g., at τ_i, the right-shift operation is first applied to the existing schedule to keep its feasibility, and second, tabu search is applied to explore better revised schedules. In the tabu search procedure,

Table 6.7 Simulation results of makespan and L_{max} for the 120 instances with $Q = 1.3$ (reprinted by permission from Elsevier [5])

	(i) Predictive makespan C_{max}									
	$Q = 1.3, R = 0.1$					$Q = 1.3, R = 0.3$				
	Index X					Index X				
(m, n)	1	2	3	4	5	1	2	3	4	5
(4, 60)	553	554	572	535	554	534	558	522	555	533
(4, 90)	839	858	823	822	851	834	821	758	792	760
(4, 120)	1059	1089	1150	1090	1084	1109	1097	1085	1091	1107
(8, 60)	584	619	641	644	632	625	621	610	631	651
(8, 90)	948	906	916	931	943	904	925	902	981	877
(8, 120)	1316	1304	1286	1270	1320	1177	1288	1276	1256	1301

	(ii) Maximum tardiness L_{max} of predictive schedule									
	$Q = 1.3, R = 0.1$					$Q = 1.3, R = 0.3$				
	Index X					Index X				
(m, n)	1	2	3	4	5	1	2	3	4	5
(4, 60)	0	0	0	0	0	17	36	2	75	48
(4, 90)	0	0	0	5	0	152	115	81	141	105
(4, 120)	0	0	0	0	0	164	196	195	186	220
(8, 60)	0	65	42	27	57	135	133	163	152	123
(8, 90)	87	74	69	97	72	236	212	195	273	186
(8, 120)	204	158	131	150	169	310	434	424	384	384

	(iii) Actual L_{max} with right-shift operation (average and worse case)									
	$Q = 1.3, R = 0.1$					$Q = 1.3, R = 0.3$				
	Index X					Index X				
(m, n)	1	2	3	4	5	1	2	3	4	5
(4, 60)	41.8	26.4	35.6	21.3	48.2	63.1	83.9	36.7	117.1	87.1
	148	87	136	108	131	178	134	128	176	182
(4, 90)	31.7	42.6	27.5	48.1	41.1	203.2	158.8	124.8	181.4	140.9
	93	157	129	133	96	293	230	218	291	210
(4, 120)	0.0	31.4	35.4	21.8	16.4	200.4	239.1	234.1	239.1	257.8
	0	136	185	77	80	275	323	336	396	317
(8, 60)	40.1	110.0	97.4	83.1	119.2	190.6	179.1	216.6	210.6	169.9
	108	223	204	158	220	283	268	310	311	270
(8, 90)	138.3	134.2	116.6	148.3	136.0	275.8	252.3	245.1	326.7	234.3
	237	240	179	281	217	321	331	358	422	312
(8, 120)	251.8	211.3	177.8	197.1	217.7	357.2	479.3	474.4	438.1	433.2
	373	355	241	311	323	511	612	615	533	544

the upper bound for the number of iterations is set to 20 as a stop criterion to reflect a quick response to disruption in dynamic environments. This is also because we could not generate any better schedule in most of the fundamental preliminary experiments even if the number of iterations exceeded 20.

When the tab search procedure detects a better revised schedule than the simple right-shift operation, it is called an *effective revision*. On the contrary, any better

Table 6.8 Simulation results of makespan and L_{max} for the 120 instances with $Q = 1.7$ (reprinted by permission from Elsevier [5])

		(i) Predictive makespan C_{max}								
		$Q = 1.7, R = 0.1$					$Q = 1.7, R = 0.3$			
		Index X					Index X			
(m, n)	1	2	3	4	5	1	2	3	4	5
$(4, 60)$	533	559	544	586	568	603	549	579	587	561
$(4, 90)$	807	891	840	880	760	804	801	870	877	806
$(4, 120)$	1236	1139	1113	1039	1142	1256	1121	1122	1199	1222
$(8, 60)$	652	696	631	630	671	638	623	641	581	612
$(8, 90)$	987	1102	979	1115	1103	917	923	917	901	932
$(8, 120)$	1382	1467	1372	1436	1447	1268	1259	1268	1307	1282

		(ii) Maximum tardiness L_{max} of predictive schedule								
		$Q = 1.7, R = 0.1$					$Q = 1.7, R = 0.3$			
		Index X					Index X			
(m, n)	1	2	3	4	5	1	2	3	4	5
$(4, 60)$	0	0	0	0	0	0	0	0	0	0
$(4, 90)$	0	0	0	0	0	0	0	0	0	0
$(4, 120)$	0	0	0	0	0	0	0	0	0	0
$(8, 60)$	0	0	0	0	0	0	0	7	0	0
$(8, 90)$	0	0	0	0	0	6	34	64	4	33
$(8, 120)$	0	0	0	0	0	103	114	79	129	161

		(iii) Actual L_{max} with right-shift operation (average and worse case)								
		$Q = 1.7, R = 0.1$					$Q = 1.7, R = 0.3$			
		Index X					Index X			
(m, n)	1	2	3	4	5	1	2	3	4	5
$(4, 60)$	0.0	0.0	0.0	0.0	0.0	26.8	19.6	13.6	31.7	19.3
	0	0	0	0	0	116	111	73	110	84
$(4, 90)$	0.0	0.0	0.0	0.0	0.0	11.5	0.0	18.1	3.0	5.3
	0	0	0	0	0	60	0	123	22	32
$(4, 120)$	0.0	0.0	0.0	0.0	0.0	39.1	0.0	12.0	3.6	20.3
	0	0	0	0	0	116	0	76	58	122
$(8, 60)$	3.4	18.5	1.1	2.4	10.9	48.7	38.8	70.4	38.1	57.3
	34	96	17	35	89	119	141	181	115	159
$(8, 90)$	5.9	24.5	5.3	9.8	51.2	61.4	88.4	111.1	49.1	94.8
	58	124	57	100	150	168	225	186	130	196
$(8, 120)$	16.3	58.3	19.7	47.4	40.0	152.4	167.3	137.2	172.4	212.3
	81	223	98	169	140	250	248	241	235	334

schedule is not found by tabu search, it is called an *ineffective revision* and then we accept the schedule by the right-shift operation.

Tables 6.7 and 6.8 show the makespan, C_{max}, and the maximum tardiness, L_{max}, of each predictive schedule before we apply the machine breakdown scenarios. Note in these tables that the instances with $(Q, R) = (1.3, 0.3)$ reflects a situation where the initial predictive schedules consist of many demanding jobs since all the instances indicate positive values of L_{max}.

These tables also show the results after we applied the 16 machine breakdown scenarios to each one of the predictive schedules. "Actual $L_{\max}$ with right-shift operation" in these tables signify the average of the maximum tardiness computed over the 16 machine breakdown scenarios together with the worst one among them when only right-shift operations were employed.

6.4.3 Pareto Optimality

In this subsection, we compare the D^*-driven scheduling policy with the EDR from the perspective of Pareto optimality. This is because we can intuitively foresee that the EDR tends to provide smaller maximum tardiness with more schedule revisions than the D^*-driven policy. Since we have introduced the maximum tardiness and the schedule revision frequency as measures of evaluation, the comprehensive analysis should be conducted on the Pareto Optimality basis.

As mentioned before, both the D^*-driven policy and the EDR were individually applied to each of the 1920 instance-scenario combinations. For the purpose of analyzing the results, let us denote, by $L_{\eta,\xi}(D^*, \tau)$, and $F_{\eta,\xi}(D^*, \tau)$, respectively, the maximum tardiness and the number of schedule revisions when we use (D^*, τ) under the D^*-driven policy in instance $\eta \in \mathcal{I}$ and machine breakdown scenario $\xi \in \mathcal{B}$. Besides, let $L_{\eta,\xi}$ and $F_{\eta,\xi}$, respectively, express the maximum tardiness and the number of schedule revisions under the EDR in instance $\eta \in \mathcal{I}$ and machine breakdown scenario $\xi \in \mathcal{B}$. Note that the notations used for the EDR do not have (D^*, τ) since the EDR is not relevant to (D^*, τ).

Based on these notations, it would be convenient to introduce three indicator variables as follows:

$$U_{\eta,\xi}^+(D^*, \tau) = \begin{cases} 1, & \text{if } L_{\eta,\xi}(D^*, \tau) \leq L_{\eta,\xi} \text{ and } F_{\eta,\xi}(D^*, \tau) \leq F_{\eta,\xi} \\ 0, & \text{otherwise.} \end{cases} \quad (6.4)$$

$$U_{\eta,\xi}^-(D^*, \tau) = \begin{cases} 1, & \text{if } L_{\eta,\xi}(D^*, \tau) \geq L_{\eta,\xi} \text{ and } F_{\eta,\xi}(D^*, \tau) \geq F_{\eta,\xi} \\ 0, & \text{otherwise.} \end{cases} \quad (6.5)$$

$$U_{\eta,\xi}(D^*, \tau) = 1 - \left[U_{\eta,\xi}^+(D^*, \tau) + U_{\eta,\xi}^-(D^*, \tau) \right], \quad (6.6)$$

where the sign of inequality must hold in at least one of the two inequality-conditions in both Eqs. (6.4) and (6.5). The indicator variable, $U_{\eta,\xi}^+(D^*, \tau)$, takes on the value 1 if the D^*-driven policy was superior to the EDR in (D^*, τ) with respect to the maximum tardiness and the schedule revision frequency, while $U_{\eta,\phi}^-(D^*, \tau)$ equals 1 if the D^*-driven policy was inferior to the EDR in (D^*, τ) with regard to the same two criteria, and moreover $U_{\eta,\phi}(D^*, \tau)$ equals 1 if the D^*-driven policy was neither superior nor inferior to the EDR in (D^*, τ). Using these three indicator variables,

Table 6.9 D^*-driven scheduling policy versus EDR—Pareto optimality: results of $NS_1(D^*, \tau)$, $NN_1(D^*, \tau)$ and $NI_1(D^*, \tau)$ (reprinted by permission from Elsevier [5])

			$D^*_{i\eta}$									
i	1				2				3			
τ	10	20	30	60	10	20	30	60	10	20	30	60
NS_1	957	973	953	929	962	927	944	899	961	954	949	914
(%)	49.8	50.7	49.6	48.4	50.1	48.3	49.2	46.8	50.1	49.7	49.4	47.6
NN_1	890	882	902	934	896	928	926	961	903	906	923	964
(%)	46.4	45.9	47.0	48.6	46.7	48.3	48.2	50.1	47.0	47.2	48.1	50.2
NI_1	73	65	65	57	62	65	50	60	56	60	48	42
(%)	3.8	3.4	3.4	3.0	3.2	3.4	2.6	3.1	2.9	3.1	2.5	2.2

D^*					$\overline{D^*}_i$							
i	1				2				3			
τ	10	20	30	60	10	20	30	60	10	20	30	60
NS_1	866	848	880	877	887	868	876	872	878	899	876	886
(%)	45.1	44.2	45.8	45.7	46.2	45.2	45.6	45.4	45.7	46.8	45.6	46.1
NN_1	1028	1043	1014	1026	1009	1024	1023	1031	1026	1001	1030	1012
(%)	53.5	54.3	52.8	53.4	52.6	53.3	53.3	53.7	53.4	52.1	53.6	52.7
NI_1	26	29	26	17	24	28	21	17	16	20	14	22
(%)	1.4	1.5	1.4	0.9	1.3	1.5	1.1	0.9	0.8	1.0	0.7	1.1

D^*					$\tilde{D}^*_i$							
i	1				2				3			
τ	10	20	30	60	10	20	30	60	10	20	30	60
NS_1	818	830	813	814	826	813	803	806	840	828	827	811
(%)	42.6	43.2	42.3	42.4	43.0	42.3	41.8	42.0	43.8	43.1	43.1	42.2
NN_1	1092	1081	1100	1096	1084	1100	1109	1106	1071	1084	1085	1106
(%)	56.9	56.3	57.3	57.1	56.5	57.3	57.8	57.6	55.8	56.5	56.5	57.6
NI_1	10	9	7	10	10	7	8	8	9	8	8	3
(%)	0.5	0.5	0.4	0.5	0.5	0.4	0.4	0.4	0.5	0.4	0.4	0.2

Table 6.10 D^*-driven scheduling policy versus EDR—sensitivity of the proposed policy in design variables (reprinted by permission from Elsevier [5])

Design variable		NS_2	(%)	NN_2	(%)	NI_2	(%)
Types of D^*	$D^*_{i\eta}$	10558	45.8	12088	52.5	394	1.7
	$\overline{D^*}_i$	10483	45.5	12197	52.9	360	1.6
	$\tilde{D}^*_i$	10623	46.1	12111	52.6	306	1.3
	$D^*_{1\eta}$	11322	49.1	11015	47.8	703	3.1
	$D^*_{2\eta}$	10513	45.6	12267	53.2	260	1.1
	$D^*_{3\eta}$	9829	42.7	13114	56.9	97	0.4
Inspection interval							
τ	10 (min)	7995	46.3	8999	52.1	286	1.7
	20 (min)	7940	45.9	9049	52.4	291	1.7
	30 (min)	7921	45.8	9112	52.7	247	1.4
	60 (min)	7808	45.2	9236	53.4	236	1.4

we will count up the number of instance-scenario combinations where each of the three indicator variables takes on the value 1 in the computational simulations.

Table 6.9 summarizes the results of this counting up operation for each (D^*, τ), where NS_1, NN_1, and NI_1 are defined by

$$NS_1(D^*, \tau) = \sum_{\eta \in \mathcal{I}} \sum_{\xi \in \mathcal{B}} U_{\eta, \xi}^+(D^*, \tau), \tag{6.7}$$

$$NN_1(D^*, \tau) = \sum_{\eta \in \mathcal{I}} \sum_{\xi \in \mathcal{B}} U_{\eta, \xi}^-(D^*, \tau), \tag{6.8}$$

$$NI_1(D^*, \tau) = \sum_{\eta \in \mathcal{I}} \sum_{\xi \in \mathcal{B}} U_{\eta, \xi}(D^*, \tau). \tag{6.9}$$

It is observed in Table 6.9 that when we used $D_{i\eta}^*$ as a critical cumulative task delay, approximately 50 % of the 1920 instance-scenario combinations show the superiority of the D^*-driven policy to the EDR, whereas only about 3 % of them indicate the inferiority of the D^*-driven policy. It is also shown in this table that when we employed $\overline{D^*}_{i\eta}$, around 45 % of the results reveal the superiority of the D^*-driven policy, while less than 2 % of them show the inferiority of the D^*-driven policy. In the case of $\tilde{D}^*$, the D^*-driven policy is superior for about 42 % of the results, and inferior for about 5 % to the EDR.

In Table 6.9, the critical cumulative delay, $D_{i\eta}^*$, indicates larger ratio than $\overline{D^*}_{i\eta}$ and $\tilde{D}_{i\eta}^*$ in reference to both the superiority and the inferiority of the D^*-driven policy. This is because $\overline{D^*}_{i\eta}$ and $\tilde{D}_{i\eta}^*$ are, respectively, the average of $D_{i\eta}^*$ over instances $\eta \in I(m, n, Q, R, *)$ and all the instances $\eta \in I(*)$; they tend to discriminate the results conservatively. It should be, however, noted that all these three kinds of the critical cumulative delays emphasize the superiority rather than the inferiority of D^*-driven policy to the EDR.

Table 6.10 summarizes the results in Table 6.9 so that the influence of the design variables, D^* and τ, upon the schedule revision frequency and the maximum tardiness becomes clearer. In this table, NS_2 signifies the number of instance-scenario combinations in which the D^*-driven policy is superior to the EDR, and NN_2 and NI_2, respectively, represent the number of combinations where the D^*-driven policy is neither superior nor inferior to the EDR and that of combinations where the D^*-driven policy is inferior to the EDR. We can notice in this table that the superiority of the D^*-driven policy to the EDR is remarkably affected neither by the type of critical cumulative delay nor by the inspection time interval, but when the critical cumulative delay is loosened, from $D_{1\eta}^*$ to $D_{2\eta}^*$ for example, its superiority will become less evident.

Table 6.11 shows the results of the sensitivity analysis in reference to parameters involved in the problem instances and the machine breakdown scenarios. In this table, NS_3 and NN_3, respectively express the number of instance-scenario combinations in which the D^*-driven policy showed its superiority to the EDR and where that was

Table 6.11 D^*-driven scheduling policy versus EDR—sensitivity of the proposed policy in reference to parameters involved in the problem or breakdown scenario (reprinted by permission from Elsevier [5])

Parameters	Values	$\mathcal{I}, \mathcal{B}$	NS_3	(%)	NN_3	(%)	NI_3	(%)
Number of machines								
m	4	$I(4, *, *, *, *), B(*)$	21171	61.3	12619	36.5	770	2.2
	8	$I(8, *, *, *, *), B(*)$	10493	30.4	23777	68.8	290	0.8
Number of jobs								
n	60	$I(*, 60, *, *, *), B(*)$	10985	47.7	11681	50.7	374	1.6
	90	$I(*, 90, *, *, *), B(*)$	10484	45.5	12210	53.0	346	1.5
	120	$I(*, 120, *, *, *), B(*)$	10195	44.2	12505	54.3	340	1.5
Due date tightness								
Q	1.3	$I(*, *, 1.3, *, *), B(*)$	8854	25.6	25214	73.0	492	1.4
	1.7	$I(*, *, 1.7, *, *), B(*)$	22810	66.0	11182	32.4	568	1.6
Due date variance								
R	0.1	$I(*, *, *, 0.1, *), B(*)$	19999	57.9	14043	40.6	518	1.5
	0.3	$I(*, *, *, 0.3, *), B(*)$	11665	33.8	22353	64.7	542	1.6
Number of machines subjected to breakdowns								
α	0.5	$I(*), B(0.5, *, *, *)$	16385	47.4	17547	50.8	628	1.8
	1.0	$I(*), B(1.0, *, *, *)$	15279	44.2	18849	54.5	432	1.3
Mean time between breakdowns								
β	0.1	$I(*), B(*, 0.1, *, *)$	13156	38.1	21389	61.9	15	0.0
	0.3	$I(*), B(*, 0.3, *, *)$	18508	53.6	15007	43.4	1045	3.0
Mean repair duration								
γ_1	0.5	$I(*), B(*, *, 0.5, *)$	15142	43.8	18701	54.1	717	2.1
	1.0	$I(*), B(*, *, 1.0, *)$	16522	47.8	17695	51.2	343	1.0
Repair duration variance								
γ_2	0.01	$I(*), B(*, *, *, 0.01)$	15859	45.9	18258	52.8	443	1.3
	0.05	$I(*), B(*, *, *, 0.05)$	15805	45.7	18138	52.5	617	1.8

neither superior nor inferior to this. It can be seen in this table that the D^*-driven policy tends to become evidently superior to the EDR in the case of $m = 4$, $Q = 1.7$, $R = 0.1$ or lower frequency of machine breakdowns. Table 6.11 also reveals that the D^*-driven policy is not inferior to the EDR even when the frequency of machine breakdowns is high, and when due dates are tight.

The above observations reveal that the D^*-driven scheduling policy is considered to be more effective than the EDR when we emphasize both the maximum tardiness and the schedule revision frequency. This is because the D^*-driven policy is more flexible than the EDR. It is not difficult to monitor task delays in the real environments, and thereby the D^*-driven policy will be one of the effective tools in the dynamic situations.

References

1. Mehta SV, Uzsoy RM (1998) Predictable scheduling of a job shop subject to breakdowns. IEEE Trans Robotics Autom 14(3):365–378
2. Morton TE, Pentico DW (1993) Heuristic scheduling systems. Wiley, New York
3. Pinedo M (2008) Scheduling—theory, algorithms, and systems, 3rd edn. Springer, New York
4. Suwa H, Sandoh H (2003) Reactive scheduling based on control limit policy with delayed tasks measurement. Trans ISCIE 16(11):565–573
5. Suwa H, Sandoh H (2007) Capability of cumulative delay based reactive scheduling for job shops with machine breakdowns. Comput Ind Eng 53:63–78

Chapter 7
Hybrid D^*-Driven Policy

Abstract This chapter incorporates the basic concept of the D^*-driven policy into the *hybrid schedule revision policy* discussed in Chap. 3 considering multiple job families on a single machine with sequence-dependent setup times. Computational simulations are carried out with the view to investigating the performances of the hybrid D^*-driven policy. Its capability and stability are also discussed through its comparison with the two types of conventional schedule revision policies; the periodic and the event-driven schedule revision policy.

7.1 Framework

In Chap. 6, we have described the D^*-driven policy as one of the reactive scheduling policies. Its properties and performances were also investigated through computational simulations in a general job shop environment to show its flexibility.In this chapter, we incorporate the underlying concept of the D^*-driven policy into the hybrid schedule revision policy that was discussed as an advanced policy in Sect. 3.5. Multiple families of jobs are considered on a single machine with sequence-dependent setup times. This is because a single machine environment will be able to promote the understanding of the hybrid D^*-driven policy due to its simple structure. Note that we do not have tasks but jobs only in a single machine environment, accordingly, the cumulative delay means the cumulative job delays. In addition, the D^*-driven policy is a specific type of event-driven policy as discussed in Sect. 5.4, and the hybrid D^*-driven policy is, likewise, a specific type of hybrid policy.

The framework of the hybrid D^*-driven policy considered here is as follows:

(1) Inspections are executed to the existing schedule at planned times $\tau_i = i\tau(\tau > 0,$ $i = 1, 2, \ldots, M)$ over a period $(0, H]$ with a view to monitoring the progress of the schedule and detecting delays, where M and H signify the number of inspections and the planning horizon, respectively.

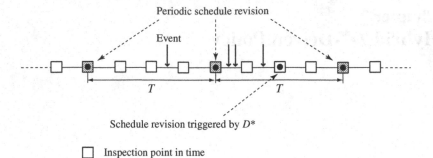

Fig. 7.1 Illustrative example of schedule revisions under the hybrid D^*-driven policy ($l = 4$, $T = l\tau = 4\tau$): a schedule revision is invoked according to (1) an amount of cumulative delay and (2) at intervals of T

(2) When the cumulative delay, D_i, at τ_i is found to have exceeded the prespecified critical cumulative delay, D^*, we conduct a schedule revision for a set of the unfinished jobs at τ_i. Let us, by S_i^P, denote the resulting new schedule for the unfinished jobs.

(3) Apart from schedule revisions triggered by the critical cumulative delay, D^*, the schedule of our concern is revised at times

$$T, 2T, \ldots, LT,$$

where $T = l\tau$, $LT \leq H$ and $(L + 1)T > H$ for $l = 1, 2, \ldots, L$.

Figure 7.1 shows the behavior of the hybrid D^*-driven policy. In this figure, inspection points in time where a schedule revision is carried out are indicated by large dots in the square on the time axis. The large dot in a shaded square expresses a periodic schedule revision, whereas the one in a non-shaded square reflects the schedule revision triggered by D^*.

7.2 Multiple Job Families on a Single Machine

7.2.1 Description of the Problem

We address multiple families of jobs on a single machine with sequence-dependent setup times and random disruptions or interruptions.

Let Y denote the number of job families, and N the number of the jobs all of which are to be processed over the planning period, $[0, H]$. Jobs in family y ($\in \{1, 2, \ldots, Y\}$) have an identical processing time given by p_y. When a job in family y' ($\in \{1, 2, \ldots, Y\}$) follows a job in family y on the machine, a sequence-dependent setup time $s_{yy'}$ (>0, $y \neq y'$) is incurred [1, 2] with $s_{yy} = 0$. Under these conditions, it is trivial that the minimum makespan is realized by minimizing the total setup times.

Let TS express the total setup times, and then it is given by

$$TS = \sum u_{jj'} s_{yy'} \quad \forall j \text{ (in } y), \; j' \text{(in } y') \; (y \neq y'), \tag{7.1}$$

where $u_{jj'}$ is an indicator variable defined by

$$u_{jj'} = \begin{cases} 1, & \text{if job } j' \text{ in } y' \text{ follows job } j \text{ in } y(y \neq y'), \\ 0, & \text{otherwise.} \end{cases} \tag{7.2}$$

7.2.2 Problem Instances

Suppose that the manufacturing environment will dynamically change due to the uncertainties discussed in Sect. 3.1. Under such an environment, we here consider the two problem instances as summarized in Table 7.1. Their details are described as follows [3, 4]:

Instance 1:   Table 7.1a summarizes the scheme of Instance 1. As shown in this table, the number, Y, of job families is 10. The length, H, of planning horizon is 8000. A total of 1000 jobs are to be processed during the planning horizon; 100 jobs among them are ready to be processed at time zero, while the other 900 jobs will randomly arrive at the system during the planning horizon.

The processing times, p_y, of jobs in family y are, regardless of family, independent and identically distributed (*iid*) random variables having the discrete uniform distribution over an interval [1, 10], which is denoted by Uniform(1, 10). The setup time, $s_{yy'}$, incurred when a job in a family y' is processed immediately after a job in a different family y follows Uniform(5, 10) regardless of job families. Besides, the inter-arrival times of jobs in y ($\in \{1, 2, \ldots, 10\}$) are *iid* random variables having the exponential distribution with arrival rate λ_y, which is denoted by EXP(λ_y) in the following. It should be noted in the above that only the parameter of the inter-arrival time distribution is relevant to the job family for simplicity.

Instance 1 reflects a situation in which machine breakdowns and spontaneous processing delays of jobs occur randomly during the schedule execution. It is assumed that the inter-breakdown times are *iid* random variables having EXP(1/1000) with rate 1/1000, i.e., with rate 0.001, while the repair times independently follow Uniform(10, 20). Further, we assume that job processing is spontaneously delayed for a certain unit of time, and is *iid* random variable with EXP(1).

Table 7.1 Simulation schemes for generating problem instances and disruptions or interruptions (reprinted by permissions from ISCIE [3] and Elsevier [4])

(a) Instance 1	
Parameters	Values
Planning horizon (H)	8000
Job family (Y)	5 ($y = 1, 2, \ldots, 10$)
Total jobs (N)	1000
Processing time (p_y)	Uniform(1, 10)
Setup time ($s_{yy'}$)	Uniform(5, 10)
Inter-arrival times	EXP(λ_y) (for 900 jobs)
Arrival rate	($\lambda_1, \ldots, \lambda_{10}$) = (0.18, 0.15, 0.13, 0.12, 0.10, 0.08, 0.07, 0.06, 0.04, 0.04)
Inter-breakdown times	EXP(1/1000)
Time to repair	Uniform(10, 20)
Processing delay	EXP(1) for each job
(b) Instance 2	
Parameters	Values
Planning horizon (H)	7000
Job family (Y)	5 ($y = 1, 2, \ldots, 5$)
Total jobs (N)	1000
Processing time (p_y)	Uniform(1, 10)
Setup time ($s_{yy'}$)	Uniform(5, 10)
Arrival interval	EXP(λ_y)
Arrival rate	($\lambda_1, \ldots, \lambda_5$) = (0.05, 0.04, 0.03, 0.02, 0.01)
Job family of urgent jobs	Uniform(1, 5)
Inter-arrival times of urgent jobs	EXP(λ)
Arrival rate	$\lambda = 0.02, 0.01, 0.005, 0.0025$

Instance 2: Table 7.1b shows the scheme of Instance 2. The number Y of job families is five, and 1000 jobs are to be processed within the period [0, 7000] ($H = 7000$). It is assumed that the inter-arrival times of jobs in family y ($\in \{1, 2, \ldots, 5\}$) are *iid* random variables having EXP(λ_y), and their processing times of jobs independently follow Uniform(1, 10). The setup time is an *iid* random variable with Uniform(5, 10). Note that the differences from Instance 1 are in the number of job families and their arrival rates.

In Instance 2, we consider a situation where, as interruptions, urgent jobs that must be processed promptly on their arrivals randomly occur during the schedule execution. Every urgent job belongs to any of the five normal job families and the probability that an urgent job in family y ($y = 1, 2, \ldots, 5$) arrives is assumed to be 1/5. Moreover, the inter-arrival times of urgent jobs in any family independently follow EXP(λ). Table 7.1b shows the values of λ used in the computational experiments. The processing times of the urgent jobs are all *iid* random variables having Uniform(1, 10).

7.3 Configuration of Scheduling Policy

7.3.1 Procedure

The procedure of the hybrid D^*-driven policy for multiple job families on a single machine is as follows:

Step 1 $\tau_0 = 0, i \leftarrow 1, \kappa \leftarrow 1$.

Step 2 If i agrees with κl, where $l = T/\tau$, then go to **Step 4** with $\kappa \leftarrow \kappa + 1$.
Else go to **Step 3**.

Step 3 Compute the total delay, $\hat{D}_i$, and then the cumulative delay, D_i, at τ_i.
If $D_i \geq D^*$, then go to **Step 4**.
Else go to **Step 5**.

Step 4 Generate a revised schedule, S_i^P, for unfinished jobs from scratch, and let $D_i = 0$.

Step 5 If $i > M$ then stop. Else $i \leftarrow i + 1$, then go to **Step 2**.

The cumulative delay, D_i, and the total delay, $\hat{D}_i$, at τ_i in **Step 3** are given by

$$D_i = \sum_{k=i'+1}^{i} \hat{D}_k, \tag{7.3}$$

$$\hat{D}_i = \begin{cases} 0, & \text{if } J_{[i]} = \phi \\ \sum_{j \in J_{[i]}} \delta_j^i, & \text{otherwise} \end{cases}, \tag{7.4}$$

where i' is defined by the relationship, $\tau_{[i]} = i'\tau$, and

$$J_{[i]} = \{j | C_j(S_{[i]}^P) \leq \tau_i, C_j(S_i^A) > \tau_{i-1}\}, \tag{7.5}$$

$$\delta_j^i = \begin{cases} \tau \ (= \tau_i - \tau_{i-1}), & \text{if } C_j(S_{[i]}^P) \leq \tau_{i-1} \text{ and } C_j(S_i^A) > \tau_i \\ \tau_i - C_j(S_{[i]}^P), & \text{if } C_j(S_{[i]}^P) > \tau_{i-1} \text{ and } C_j(S_i^A) > \tau_i \\ C_j(S_i^A) - \tau_{i-1}, & \text{if } C_j(S_{[i]}^P) \leq \tau_{i-1} \text{ and } C_j(S_i^A) \leq \tau_i \\ C_j(S_i^A) - C_j(S_{[i]}^P), & \text{if } C_j(S_{[i]}^P) > \tau_{i-1} \text{ and } C_j(S_i^A) \leq \tau_i \end{cases} \tag{7.6}$$

Note that, after a schedule revision is carried out at $\tau_{[i]}$, the total delays of jobs on schedule, $S_{[i]}^P$, are reset to zero, that is,

$$\hat{D}_{[i]} = 0. \tag{7.7}$$

7.3.2 Scheduling Method in a Dynamic Environment

In a dynamic single machine environment, one of our major concerns is how to revise the existing schedule for the jobs in the queue waiting for being processed by the machine. We here employ the method by Vieira [5] whose algorithm can be described as follows:

Step 1 Append the newly-arrived jobs to the queue, and resequence all the jobs in the queue so that jobs in a family are in the same group. This is to reduce the total setup times.

Step 2 In each group, sequence the jobs by FIFO.

Step 3 Detect the job, j, that has just been processed or is still being processed in the actual schedule, S_i^A, and then let the job group belonging to the same family as j be processed first in the revised schedule, S_i^P.

Step 4 Sequence the remaining job groups by FIFO with regard to the first job in each group.

It is postulated that urgent jobs have a tight deadline, accordingly, they have higher priority than the normal jobs in the queue; the urgent jobs are to be processed prior to the other jobs but immediately after the job being processed currently. This will cause an increase in the total setup times to decrease the productivity.

7.4 Computational Experiments

7.4.1 Design Variables

In this section, we perform computational experiments of the hybrid D^*-driven policy in order to investigate its characteristics. The hybrid D^*-driven policy has three design variables, the critical cumulative delay, D^*, the schedule revision interval, T, and the inspection interval τ. We here introduce a set of candidate values for these three design variables since it is very difficult to find their optimal values analytically.

The candidate set of the design variables considered in each problem instance is as follows:

Instance 1:

$$\tau = 10, 100,$$
$$D^* = 250, 500, 750, \ldots, 10000 \text{ (a total of 40 candidates)},$$
$$T = 400, 800.$$

As a result, Instance 1 yields 160 cases to be examined.

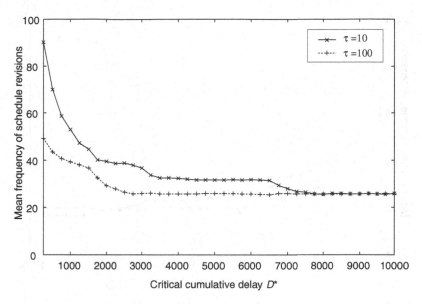

Fig. 7.2 Mean frequency of schedule revisions in the case of $T = 400$ (reprinted with kind permission from ISCIE [3])

Instance 2:

$$\tau = 10,$$
$$D^* = 500, 6000, 10000,$$
$$T = 200, 300, 400, 500, 600, 700, 800, 900, 1000.$$

The above candidate set of design variables provides 27 cases in Instance 2.

7.4.2 Behaviors

Instance 1

In Instance 1, we generated 10 scheduling problem instances according to the scheme summarized in Table 7.1a. In addition, we applied each one of the 160 cases of the design variables to each individual problem instance to investigate the behaviors of the hybrid D^*-driven policy. Figures 7.2 and 7.3 show the simulation results in the case of $T = 400$ and 800, respectively. They reveal the behavior of the average number of schedule revisions, that is, the mean frequency of schedule revisions, triggered by D^* over the 10 problem instances against the critical cumulative delay, D^*. It is observed in both figures that the mean frequency of schedule revisions by

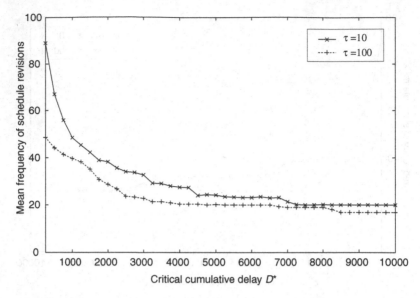

Fig. 7.3 Mean frequency of schedule revisions in the case of $T = 800$ (reprinted with kind permission from ISCIE [3])

D^* decreases with increasing D^*. This is because a periodic schedule revision will be carried out before the cumulative delay reaches D^* for a large value of D^*.

On the other hand, Figs. 7.4 and 7.5 depict the mean total setup times over the ten problem instances in the case of $T = 400$ and 800, respectively. Figure 7.4 indicates the mean total setup times gradually increase with D^* for both $\tau = 10$ and 100, whereas that in Fig. 7.5 shows a quite different behavior for $\tau = 100$ from that in Fig. 7.4. The mean total setup times remarkably increases when D^* exceeds around 6300 for $\tau = 100$, suggesting that we should be careful when both τ and D^* take on large values.

The hybrid policy with $\tau = 100$ results in lower frequency of schedule revisions than the hybrid policy with $\tau = 100$. Conversely, the hybrid policy with $\tau = 100$ can generate better schedules than the hybrid policy $\tau = 10$ in the sense of minimizing the total setup times. This tendency indicates that there exists a trade-off between the frequency of schedule revision and the total setup times with regard to the inspection interval.

Instance 2

In Instance 2, we generated 50 scheduling problem instances for each arrival rate of urgent jobs in Table 7.1b to yield a total of 200 problem instances, and then applied each one of the 27 cases of the design variables to each individual problem instance.

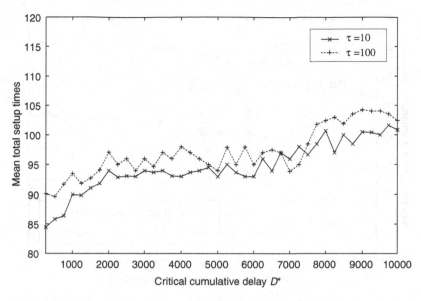

Fig. 7.4 Mean total setup times in the case of $T = 400$ (reprinted with kind permission from ISCIE [3])

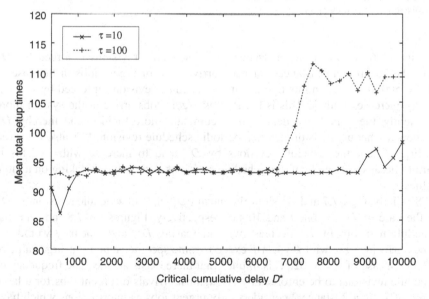

Fig. 7.5 Mean total setup times in the case of $T = 800$ (reprinted with kind permission from ISCIE [3])

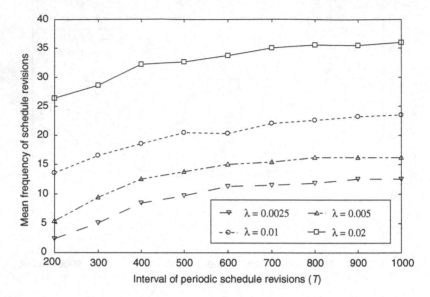

Fig. 7.6 Simulation results of mean frequency of schedule revisions with $D^* = 500$ (**Instance 2**): In the case of $T = 200$, schedule revisions invoked by T were conducted from 33 to 36 times on average. As the interval T increases, the frequency of schedule revisions by T becomes lower while the execution of D^*-driven schedule revision becomes more frequent (reprinted with kind permission from Elsevier [4])

Figure 7.6 shows the mean frequency of schedule revisions enforced by D^* over the 50 problem instances for each arrival rate of urgent jobs in the case of $D^* = 500$. It is seen in this figure that the schedule revisions enforced by D^* evidently increases with λ. This is because as urgent jobs arrive at the system more frequently, the cumulative delay will accumulate more rapidly, and thereby D^* triggers a schedule revision before a periodic schedule revision. It is also observed in Fig. 7.6 that the schedule revisions by D^* tend to increase with T. This is simply because D^* will enforce schedule revisions earlier than T taking on a large value.

Similarly, Figs. 7.7 and 7.8 show the mean frequency of schedule revision by D^* in the case of $D^* = 6000$ and 10000, respectively. Figures 7.6–7.8 indicate that schedule revisions by D^* decrease with increasing D^*, and that they would not necessarily increase with T when T exceeds some specific value for a large value of D^* in the case of $\lambda = 0.02$. This implies that there exists an adequate frequency of schedule revisions to be enforced by D^* against arrivals of urgent jobs for a large value of T since Instance 2 considers only urgent jobs as interruptions which have control over behaviors of the cumulative delays.

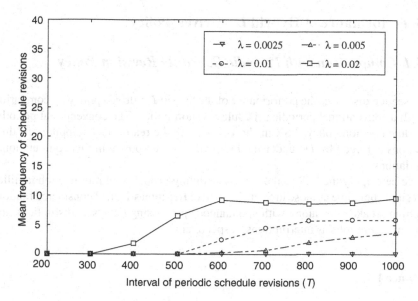

Fig. 7.7 Simulation results of mean frequency of schedule revisions with $D^* = 6000$ (**Instance 2**): The frequency of D^*-based schedule revision becomes less than that of $D^* = 500$. In the case of $\lambda = 0.0025$, schedule revision invoked by T was only performed (reprinted with kind permission from Elsevier [4])

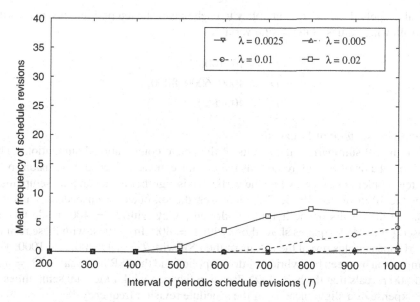

Fig. 7.8 Simulation results of mean frequency of schedule revisions with $D^* = 10000$ (**Instance 2**): The frequency of D^*-based schedule revision tends to be much less than those by other D^* values because the possibility that the critical cumulative delay exceeds D^* becomes lower (reprinted with kind permission from Elsevier [4])

7.5 Performance of Hybrid D^*-Driven Policy

7.5.1 Comparison with Periodic Schedule Revision Policy

This section discusses the performance of the hybrid D^*-driven policy by comparing it with a conventional periodic schedule revision policy. The conventional periodic schedule revision policy (PSR in short) can easily be realized by skipping schedule revisions enforced by D^* under the hybrid D^*-driven policy in this computational simulations.

We here apply the PSR to the same scheduling problems of multiple job families on a single machine as those in Sect. 7.4, where **Instances 1** and **2** considered random machine breakdowns along with spontaneous processing delays as disruptions and random urgent jobs as interruptions, respectively.

Instance 1

To each one of the ten problem instances generated in Instance 1 in Sect. 7.4.2, we applied the PSR with $\tau = 10$ and

$$T = 100, 200, \ldots, 1000 \text{ (a total of 10 candidates)}.$$

We also applied the D^*-driven policy to each one of the ten problem instances with the following values of the design variables:

$$\begin{aligned}
\tau &= 10, 100, \\
D^* &= 4000, 6000, 8000, \\
T &= 400, 800,
\end{aligned}$$

which yields a total of 12 cases.

Figure 7.9 summarizes the results of the above computational simulations. The horizontal axis of Fig. 7.9 represents the mean frequency of schedule revisions over the ten problem instances, while the vertical axis signifies the mean total setup times over the 10 instances. In Fig. 7.9, moreover, the six open (or unshaded) diamonds indicate the results of the hybrid D^*-driven policy with $T = 400$, while the six closed (or shaded) squares show those with $T = 800$. In contrast with these results, the 10 open circles reveal the results of the PSR with $T = 100, 200, \ldots, 1000$. The comparison between the hybrid D^*-driven policy and the PSR in the case of $T = 400$ and 800 reveals that the hybrid D^*-driven policy can inhibit the total setup times at the expense of a slight increase of the schedule revision frequency.

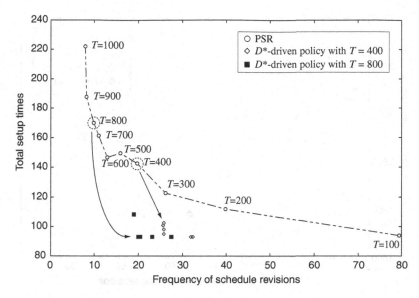

Fig. 7.9 Hybrid D^*-driven policy versus periodic schedule revision policy (reprinted with kind permission from ISCIE [3])

Instance 2

To each one of the 50 problem instances generated for each arrival rate of urgent jobs in Sect. 7.4.2, we here applied the PSR with $\tau = 10$ and

$$T = 200, 600, 1000.$$

To the same problem instances, we also applied the hybrid D^*-driven policy with

$$
\begin{aligned}
\tau &= 10, \\
D^* &= 500, 600, 10000, \\
T &= 200, 600, 1000,
\end{aligned}
$$

which yields a total of nine cases.

Figures 7.10–7.13 summarize the results of the computational simulations. In these figures, the horizontal axis represents the mean frequency of schedule revisions over the 50 problem instances for each arrival rate of urgent jobs, and the vertical one expresses the mean total setup times. In addition, the three open squares indicate the results of the hybrid D^*-driven policy with $T = 200, 600, 1000$ in the case of $D^* = 500$, and the three open circles and triangles show those when we have $D^* = 600$ and $D^* = 10000$, respectively. The open diamonds represent the results of the PSR with $T = 200, 600, 1000$.

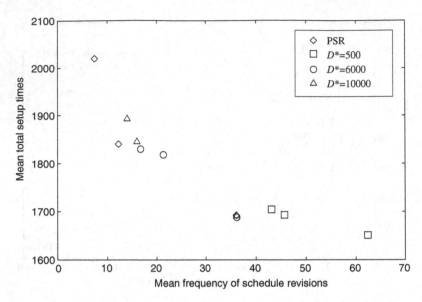

Fig. 7.10 PSR/hybrid D^*-driven policy with $\lambda = 0.02$ (reprinted with kind permission from Elsevier [4])

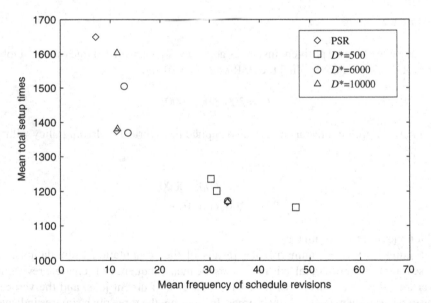

Fig. 7.11 PSR versus hybrid D^*-driven policy with $\lambda = 0.01$ (reprinted with kind permission from Elsevier [4])

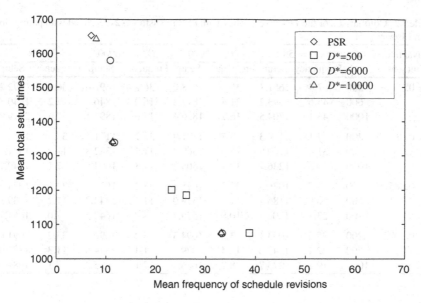

Fig. 7.12 PSR versus hybrid D^*-driven policy with $\lambda = 0.005$ (reprinted with kind permission from Elsevier [4])

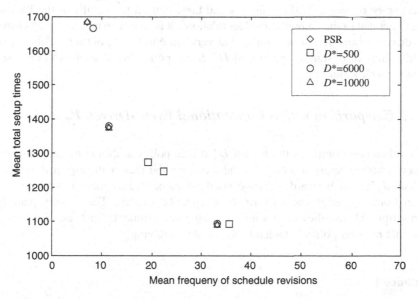

Fig. 7.13 PSR versus hybrid D^*-driven policy with $\lambda = 0.025$ (reprinted with kind permission from Elsevier [4])

Table 7.2 Comparison of the D^*-driven policy with PSR in **Instance 2** (reprinted with permission from Elsevier [4])

Arrival rate of urgent jobs	T	$D^* = 500$		$D^* = 6000$		$D^* = 10000$		PSR	
		Frequency	Setup	Frequency	Setup	Frequency	Setup	Frequency	Setup
$\lambda = 0.02$	200	62.4	1651.1	36.2	1688.0	36.2	1692.6	36.1	1692.2
	600	45.6	1693.2	21.4	1817.8	16.0	1846.9	12.2	1840.8
	1000	43.1	1704.5	16.7	1830.4	14.0	1894.1	7.4	2019.8
$\lambda = 0.01$	200	47.1	1154.3	33.7	1171.9	33.7	1171.4	33.6	1168.3
	600	31.4	1200.9	13.5	1369.2	11.4	1382.8	11.2	1374.1
	1000	30.3	1236.7	12.7	1505.2	11.3	1604.6	7.1	1649.7
$\lambda = 0.005$	200	38.7	1076.3	33.3	1075.1	33.2	1077.1	33.1	1074.3
	600	26.0	1186.7	11.6	1339.0	11.3	1343.2	11.2	1339.5
	1000	23.0	1202.8	10.8	1579.2	8.0	1642.5	7.0	1651.74
$\lambda = 0.0025$	200	35.6	1093.1	33.2	1094.3	33.2	1090.6	33.2	1091.6
	600	22.4	1247.0	11.4	1379.3	11.4	1374.8	11.3	1375.0
	1000	19.2	1273.7	8.2	1666.3	7.1	1685.3	7.1	1684.6

Table 7.2 shows the detailed information. Note in this table that the hybrid D^*-driven policy with $D^* = 10000$ shows results similar to those of the PSR since most of the schedule revisions carried out are periodic ones for a large value of D^*.

It can be observed in these figures and table that all the results of the PSR are Pareto optimal in the computational simulations. It is also seen that the results by the D^*-driven policy are not necessarily but very close to Pareto optimal. This implies the difficulty of designing the hybrid D^*-driven policy because of its three design variables D^*, T, and τ.

7.5.2 Comparison with a Conventional Event-Driven Policy

This subsection compares the hybrid D^*-driven policy with a conventional event-driven schedule revision policy. We here employ the idea with respect to an event by Vieira [5], which regards the time when the queue size reaches to its prespecified upper bound q_s as an event to trigger a schedule revision. This is very plausible in multiple job families on a single machine environment. Such an event-driven schedule revision policy is called EDQS in the following.

Instance 1

To each one of the 10 problem instances generated in Instance 1 in Sect. 7.4.2, we applied the EDQS with $\tau = 10$ and

$$q_s = 4, 8, 16, \ldots, 80, \quad \text{a total of 20 candidates.}$$

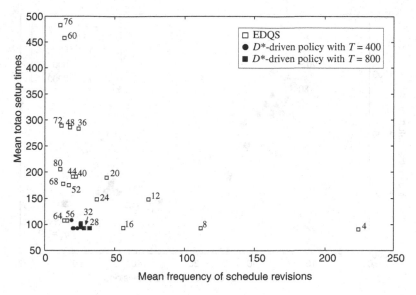

Fig. 7.14 Comparison between hybrid D^*-driven policy and EDQS (reprinted with kind permission from ISCIE [3])

Similarly, we also applied the hybrid D^*-driven policy to the same problem instances with the design variables set to:

$$\tau = 10, 100,$$
$$D^* = 4000, 6000, 8000,$$
$$T = 400, 800.$$

Figure 7.14 depicts the results of the computational simulations. The horizontal and the vertical axes of Fig. 7.14, respectively, represent the mean frequency of schedule revisions and the mean total setup times over the 10 instance problems. In this figure, the 20 open squares represent the results of the EDQS; the values of q_s are indicated beside individual square. Although we have difficulty in identifying symbols due to overlapping, the six closed circles and squares represent the results of the hybrid D^*-driven policy with $T = 400$ and 800, respectively.

It can be seen in this figure that the results of the EDQS vary widely depending on the value of q_s. When q_s takes a small value, the mean total setup times tend to become small due to frequent schedule revisions. When q_s is large, conversely, the mean total setup times tend to be large since schedule revisions will be carried out less frequently. However, note that most of them are not Pareto optimal. This is because the EDQS focuses on the queue length although neither machine breakdowns nor spontaneous processing delays, which are the causes of delays in Instance 1, affect the queue length. In contrast, the results of the hybrid D^*-driven policy show its steadiness since most of its results are Pareto optimal or very close.

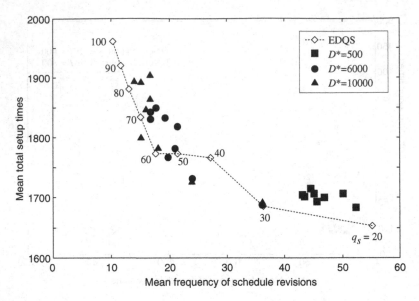

Fig. 7.15 EDQS versus hybrid D^*-driven policy with $\lambda = 0.02$: The hybrid D^*-driven policy with $T = 300$ (and $D^* = 6000$ or 10000) provides better schedules while it provides no good schedules when $D^* = 500$ (reprinted with kind permission from Elsevier [4])

Instance 2

We applied the EDQS to the 50 problem instances generated for each arrival rate of urgent jobs in Instance 2 in Sect. 7.4.2. The values of the design variables are $\tau = 10$ and

$$q_s = 20, 30, 40, 50, 60, 70, 80, 90, 100,$$

which yields nine cases. We also applied the hybrid D^*-driven policy to the same problem instances with

$$\begin{aligned}
\tau &= 10, \\
D^* &= 500, 600, 10000, \\
T &= 200, 300, 400, \ldots, 1000,
\end{aligned}$$

which yields 27 cases.

Figures 7.15–7.18 summarize the results of the computational experiments. The horizontal and the vertical axes of these figures, respectively, represent the mean frequency of schedule revisions and the mean total setup times over 50 instances for each arrival rate of urgent jobs. In these figures, the nine open diamonds show the results of the EDQS; the values of q_s are indicated beside the individual diamonds. The nine closed squares, circles, and triangles correspond to $D^* = 500$, 6000, and 10000, respectively.

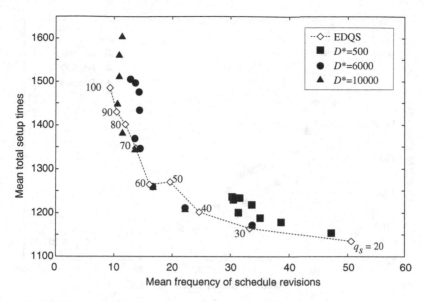

Fig. 7.16 EDQS versus hybrid D^*-driven policy with $\lambda = 0.01$: The tendency is almost the same as that with $\lambda = 0.02$. The combination of $D^* = 6000$ and $T = 300$ seems to be suitable for reducing total setups as well as frequency of schedule revisions (reprinted with kind permission from Elsevier [4])

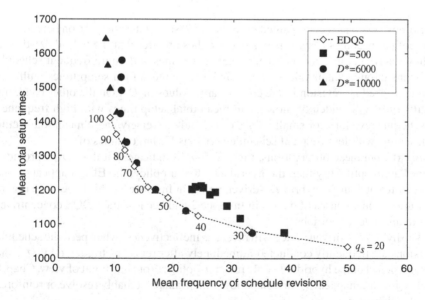

Fig. 7.17 EDQS versus hybrid D^*-driven policy with $\lambda = 0.005$: The combination of $D^* = 6000$ and $T = 300$ is suitable (reprinted with kind permission from Elsevier [4])

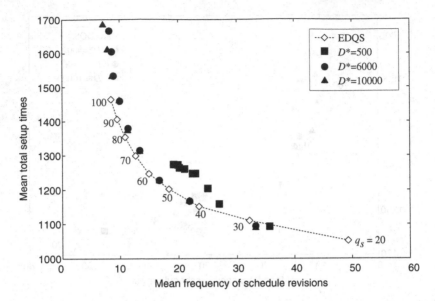

Fig. 7.18 EDQS versus hybrid D^*-driven policy with $\lambda = 0.0025$: In the case of $D^* = 500$ and $T = 200$, the hybrid D^*-driven policy outperforms EDQS (reprinted with kind permission from Elsevier [4])

Some of them are accompanied by values of T so that the comparison between the two polices becomes easy. It is observed in these figures that the hybrid D^*-driven policy tends to realize a small mean total setup times with more frequent schedule revisions through small values of D^*, and a large mean total setup times with less frequent schedule revision in the case of large values of D^*. On the other hand, the results of EDQS evidently show small mean total setup times with high frequency of schedule revisions for small values of q_s, and conversely, large mean total setup times along with less frequent schedule revisions for large values of q_s.

Apart from these observations, Figs. 7.15–7.18 indicate that the EDQS provides more Pareto optimality than the hybrid D^*-driven policy; the EDQS appears to be more efficient than the hybrid D^*-driven policy in Instance 2. This is because urgent jobs are the only causes of delays in Instance 2 and because the EDQS concentrates on the queue length of jobs.

The hybrid D^*-driven policy will reveal its ineffectiveness when periodic schedule revisions are frequently conducted immediately after revisions triggered by D^*, and this is a defect of the hybrid schedule revision policy. From this point of view, Chap. 8 deals with an enhanced D^*-driven policy, which can presumably resolve or reinforce this disadvantage.

References

1. Allahverdi A, Gupta JND, Aldowaisan T (1999) A review of scheduling research involving setup considerations. Omega, Int J Manag Sci 27:219–239
2. Pinedo M (2008) Scheduling—theory, algorithms, and systems, 3rd edn. Springer, New York
3. Suwa H, Fujimura D, Sandoh H (2004) Cumulative delay based reactive scheduling policy—its application to single-machine dynamic scheduling. Trans ISCIE 17(9):371–378
4. Suwa H (2007) A new when-to-schedule policy in online scheduling based on cumulative task delays. Int J Prod Econ 110:175–186
5. Vieira GE, Herrmann JW, Lin E (2000) Analytical models to predict the performance of a single-machine system under periodic and event-driven rescheduling strategies. Int J Prod Res 38(8):1899–1915

Chapter 8
Enhanced D^*-Driven Policy

Abstract This chapter deals with the *enhanced D^*-driven policy*, which incorporates the D^*-driven policy into the *enhanced schedule revision policy* discussed in Chap. 3. We also examine the effectiveness of the enhanced D^*-driven policy, considering dynamic flexible flow shop problems with urgent jobs as interruptions. We first develop the basic idea and the framework of the enhanced D^*-driven policy, and then describe flexible flow shop scheduling problems with urgent jobs. Through a series of computational simulations, we investigate the performance of the enhanced D^*-driven policy by comparing it with various schedule revision policies to show the effectiveness of the enhanced D^*-driven policy.

8.1 Framework

As clarified in Chap. 7, the hybrid D^*-driven policy occasionally shows its ineffectiveness especially when a schedule revision triggered by D^* is carried out immediately after a periodic schedule revision. In order to resolve or reinforce this disadvantage, we can utilize the advantage of the enhanced schedule revision policy.

The enhanced D^*-driven policy performs inspections to the existing schedule in order to detect schedule delays at planned inspection points in time

$$\tau_i = i\tau \quad (\tau > 0, \ i = 1, 2, \ldots, M)$$

over a period $(0, H]$ with $M\tau \le H$. At each inspection point, τ_i, a schedule revision is carried out if the cumulative delay, D_i, exceeds a prespecified critical cumulative delay, D^*, or the elapsed time since the previous schedule revision reaches T, whichever occurs first. After the schedule revision, the current cumulative delay, D_i, is set to zero. Figure 8.1 shows the behavior of the enhanced D^*-driven policy.

H. Suwa and H. Sandoh, *Online Scheduling in Manufacturing*,
DOI: 10.1007/978-1-4471-4561-5_8, © Springer-Verlag London 2013

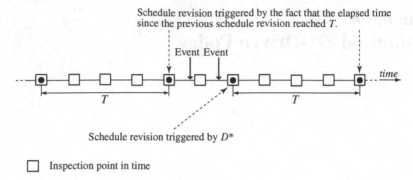

Fig. 8.1 Illustrative example of the scheduling timing under the enhanced D^*-driven policy. A schedule revision is performed if the cumulative delay, D_i, exceeds D^*, or the elapsed time since the previous schedule revision reaches T, whichever occurs first

8.2 Flexible Flow Shops

8.2.1 Definition of Flexible Flow Shop Problems

The notations used here are listed below:

c the number of stages.
n the number of jobs to be processed within the period $[0, H]$.
θ index number of stages ($\theta = 1, 2, \ldots, c$).
μ_θ the number of machines at stage θ.
H planning horizon.

In this chapter, we deal with flexible flow shop problems, where the flexible flow shop is a machine environment with c stages in series, and each stage θ ($\theta = 1, \ldots, c$) consists of μ_θ (≥ 1) identical machines in parallel. In such an environment, we consider multiple job families, which was discussed on a single machine in Chap. 7.

When a job in family y' immediately follows a job in family y on the same machine, the setup time, $s_{yy'}$ ($s_{yy'} > 0, y \neq y'$), is incurred between them. The setup time between two consecutive jobs in the same family is zero, i.e., $s_{yy} = 0$. There is unlimited work-in-process storage between two successive stages θ and $\theta + 1$ ($\theta = 1, \ldots, c - 1$). The transportation time between two successive stages can be negligible. All the jobs are to go through all the stages from stage 1 through stage c; at each stage, each individual job is to be processed by one of the identical machines. It is assumed that the release dates of some jobs are not known in advance since they arrive at the system randomly in time. Besides, interruptions occur randomly during the schedule execution, indicating the necessity of online scheduling.

Fig. 8.2 Flexible flow shop
with three stages. The symbol
R_k^θ indicates machine k at
stage θ. Stage 2 has only
one machine which may
become the bottleneck for
production flow

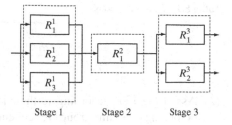

Stage 1   Stage 2   Stage 3

In the above machine environment, it is necessary to determine the job sequence
at each individual stage since the sequence-dependent setup times have influence
upon the total performances. In this chapter, we confine ourselves to the following
two criteria:

- Average completion time
- Total setup time.

8.2.2 Methods for Schedule Revision

Quadt and Kuhn [7] have presented a taxonomy of flexible flow shop scheduling
procedures. First, they categorized the procedures proposed in the literature broadly
into optimizing procedures and heuristic ones.

The branch and bound method is representative of optimizing procedures (e.g. [1]),
while the heuristic procedures are further classified into holistic and decomposition
approaches. The holistic approaches attempt to solve a given scheduling problem in
an integrated way while decomposition approaches divide an original problem into
several subproblems in many cases, which may allow a simplification of the overall
problem [2, 3, 12]. The holistic approaches include dispatching rules such as Longest
processing time first(LPT) [11], list scheduling [8], local search, and metaheuristics.
The AI-based scheduling methods [5] might also be categorized as one of the holistic
approaches. In the decomposition approaches, there are job-oriented, stage-oriented,
and problem-oriented decomposition approaches. A detailed discussion of decomposition approaches can be found in [4, 7].

In this chapter, we use the *list scheduling* [6, 9] as a method for determining the
sequence of jobs on an individual machine at each stage. The procedure of schedule
revision can be summarized as follows [10]:

(1) At an inspection point in time, τ_i, let us concentrate on the jobs in the work-
   in-process storage in front of the focal stage. Some of them were scheduled
   at an inspection point before τ_i but have not been completed by τ_i, while the
   others have arrived at the machine since the latest rescheduling but have not been
   scheduled yet.

Table 8.1 Example of jobs

Job	J_3	J_4	J_5	J_6	J_7	J_8	J_9	J_{10}
Family	3	2	1	3	2	1	3	1
Arrival time	15	10	12	8	17	22	21	19

(2) Assort the jobs according to their job families to form job groups. The jobs belonging to a same group are sequenced in FIFO manner, i.e., in each individual group, what comes in first is processed first.

(3) Assign a machine to the job group belonging to the same family as the last job which has just been processed or the job still being processed.

(4) Sequence the remaining job groups by FIFO of the first job in each group. Then machine assignment is conducted in a list scheduling manner with consideration to the setup time to be incurred.

Using the example as shown in Fig. 8.2, let us see how jobs in the storage is sequenced by the above procedure. In Fig. 8.2, the system is composed of three stages ($c = 3$). Stage 1 consists of three identical machines ($\mu_1 = 3$) and stage 3 is equipped with two identical machines ($\mu_3 = 2$), while stage 2 has only one machine ($\mu_2 = 1$).

We here concentrate on stage 3, assuming:

- At a certain inspection point, we have four jobs J_3, J_4, J_5, and J_6 that were scheduled at the previous rescheduling, but have not yet been completed;
- Machines R_1^3 and R_2^3 are processing J_1 and J_2, respectively, where J_1 and J_2 belong to families 1 and 2, respectively;
- There also exist four jobs J_7, J_8, J_9, and J_{10} that have arrived before the current scheduling point and have not yet been scheduled;
- Their families and arrival times at stage 3 are summarized in Table 8.1.

A schedule for the eight jobs at stage 3 is generated by the following steps:

(1) Group the jobs according to the family, and then, in each group, sequence the jobs based on FIFO rule. The results are $\{J_5, J_{10}, J_8\}$, $\{J_4, J_7\}$, and $\{J_6, J_3, J_9\}$.

(2) Assign the job group in family 1, $\{J_5, J_{10}, J_8\}$, to machine R_1^3. Hence, the completion time of J_8 is predictively computed. Let C_8 denote the predictive completion time of J_8; R_1^3 is predicted to become free after J_8.

(3) Allocate the job group in family 3, $\{J_6, J_3, J_9\}$, to machine R_2^3. The completion time of J_9, which is written as C_9, is predictively obtained; machine R_2^3 is predicted to be free after C_9.

(4) Allocate the remaining group $\{J_4, J_7\}$ of family 3 to either R_1^3 or R_2^3 according to the list scheduling manner; for example, allocate $\{J_4, J_7\}$ to machine R_2^3 if $C_8 + s_{12} > C_9 + s_{32}$.

As a result of the above sequencing, we have

$$R_1^3 : \{J_5, J_{10}, J_8\}$$
$$R_2^3 : \{J_6, J_3, J_9\} \rightarrow \{J_4, J_7\}$$

Note in the above that no setup time is incurred in processing the first job J_5 in family 1 as well as J_6 in family 3, however, the setup time, s_{32}, is required between J_9 and J_4.

8.3 Configuration of Enhanced Policy

8.3.1 Definition of Cumulative Delay

Let $J_{i,\theta}^A$ denote the set of jobs that have already been completed at stage θ before τ_i, and $J_{i,\theta}^P$ represent the set of unfinished jobs in the storage in front of stage θ, which are to be processed after τ_i. Further, let S_i^A represent the actually realized schedule over period $(0, \tau_i]$, then S_i^A might include some delays; the actual schedule, S_i^A, might differ from the planned one.

Let us define, by $\tau_{[i]}$, the inspection point in time before τ_i where the most recent rescheduling was conducted. In addition, let $S_{[i]}^P$ express the predictive schedule obtained by the rescheduling, and S_i^P signify the part of $S_{[i]}^P$ after τ_i ($\tau_i > \tau_{[i]}$).

On the basis of these preliminaries, we define the sets, $J_{i,\theta}^{(l)}$ ($l = 1, 2, 3, 4$), by

$$J_{i,\theta}^{(1)} = \left\{ j \,\middle|\, j \in J_{i,\theta}^P, \; \tau_{[i]} < C_{j,\theta}(S_{[i]}^P) \leq \tau_{i-1}, \right\},$$

$$J_{i,\theta}^{(2)} = \left\{ j \,\middle|\, j \in J_{i,\theta}^P, \; \tau_{i-1} < C_{j,\theta}(S_{[i]}^P) \leq \tau_i \right\},$$

$$J_{i,\theta}^{(3)} = \left\{ j \,\middle|\, \tau_{i-1} < C_{j,\theta}(S_i^A) \leq \tau_i, \; \tau_{[i]} < C_{j,\theta}(S_{[i]}^P) \leq \tau_{i-1} \right\},$$

$$J_{i,\theta}^{(4)} = \left\{ j \,\middle|\, \tau_{i-1} < C_{j,\theta}(S_i^A) \leq \tau_i, \; \tau_{i-1} < C_{j,\theta}(S_{[i]}^P) \leq \tau_i \right\},$$

where $C_{j,\theta}(s)$ denotes the completion time of job j on schedule s at stage θ. Using these notations, let us consider a set of jobs at stage θ defined by

$$\tilde{J}_{i,\theta} = J_{i,\theta}^{(1)} \cup J_{i,\theta}^{(2)} \cup J_{i,\theta}^{(3)} \cup J_{i,\theta}^{(4)}.$$

The job sets $J_{i,\theta}^{(1)}$ and $J_{i,\theta}^{(2)}$ defined in the above contain unfinished jobs at τ_i although the jobs in $J_{i,\theta}^{(1)}$ and $J_{i,\theta}^{(2)}$ were supposed to have been completed before τ_{i-1} and during the time interval $(\tau_{i-1}, \tau_i]$, respectively, in $S_{[i]}^P$. On the other hand, the job sets $J_{i,\theta}^{(3)}$ and $J_{i,\theta}^{(4)}$ contain jobs which were completed between τ_{i-1} and τ_i. The completion times of the finished jobs in $J_{i,\theta}^{(3)}$ were scheduled before τ_{i-1} in $S_{[i]}^P$, while those in $J_{i,\theta}^{(4)}$ were scheduled between τ_{i-1} and τ_i.

These observations reveal that the set $\tilde{J}_{i,\theta}$ consists of jobs which have not yet been finished by τ_{i-1} although they were supposed to have been completed before τ_i in $S_{[i]}^P$.

Now, we confine ourselves to the cumulative delay, $D_{i,\theta}$, observed at inspection point in time τ_i at stage θ. It is defined by

$$D_{i,\theta} = D_{i-1,\theta} + \sum_{j \in \tilde{J}_{i,\theta}} \delta_{i,\theta}^j \quad (i > 1), \tag{8.1}$$

where $\delta_{i,\theta}^j$ expresses a delay of job j $\left(\in \tilde{J}_{i,\theta} (\neq \phi) \right)$ over period $(\tau_{i-1}, \tau_i]$, and it is given by

$$\delta_{i,\theta}^j = \begin{cases} \tau, & \text{if } j \in J_{i,\theta}^1 \\ \tau_i - C_{j,\theta}(S_{[i]}^P), & \text{if } j \in J_{i,\theta}^2 \\ C_{j,\theta}(S_i^A) - \tau_{i-1}, & \text{if } j \in J_{i,\theta}^3 \\ 0, & \text{if } j \in J_{i,\theta}^4 \text{ and } C_{j,\theta}(S_{[i]}^P) = C_{j,\theta}(S_i^A) \\ C_{j,\theta}(S_i^A) - C_{j,\theta}(S_{[i]}^P), & \text{if } j \in J_{i,,\theta}^4 \text{ and } C_{j,\theta}(S_{[i]}^P) < C_{j,\theta}(S_i^A) \end{cases} \tag{8.2}$$

and $D_{0,\theta} \equiv 0$. When a schedule revision is conducted at τ_i, then $D_{i,\theta}$ is set to zero. The *cumulative delay*, D_i, at τ_i is then given by

$$D_i = \sum_{\theta=1}^c D_{i,\theta} \quad (i > 1). \tag{8.3}$$

8.3.2 Procedure

Suppose that we make a judgment concerning a schedule revision at each individual stage, and then the procedure of enhanced D^*-driven scheduling policy can be described as follows:

Enhanced D^*-driven scheduling policy:

Step 1 $i \leftarrow 1. q \leftarrow 0.$

Step 2 At τ_i, compute the cumulative task delay, $D_{i,\theta}$, and the elapsed time since the most recent schedule revision for $\theta = 1, 2, \ldots, c$. If $D_{i,\theta'} \geq D_{\theta'}^*$ holds or the elapsed time since the most recent rescheduling reaches T whichever occurs first at stage $\theta' \in \{1, \ldots, c\}$, revise the existing schedule for the unfinished jobs at stage θ'. The schedules for the unfinished jobs at stages $\theta = \theta' + 1, \theta' + 2, \ldots, c$ are also revised. Let $D_{i,\theta} = 0$ for $\theta = \theta', \ldots, c$ and $q \leftarrow i.$

Table 8.2 Schemes of computational experiments–generation of problem instances

Parameters	Values
Job families	5
Number of planned jobs	1,000
Processing time required for each stage	Family 1: Uniform(5,7)
	Family 2: Uniform(9,10)
	Family 3: Uniform(11,13)
	Family 4: Uniform(13,15)
	Family 5: Uniform(15,17)
Setup time	Uniform(1,5)
Arrival interval	$\text{Exp}(1/\lambda')$ where $\lambda' = 0.07$

Table 8.3 Schemes of computational experiments—occurrence of urgent jobs (1)

Parameters	Values
Job families	Uniform(1, 5)
Arrival interval	$\text{Exp}(1/\lambda)$
	where $\lambda = 0.01$

Step 3 If $\tau_i > H$ or all the jobs are completed, then stop. Else $i \leftarrow i + 1$, then go back to **Step 2**.

In **Step 2**, the right-shift operation is tentatively applied to the existing schedule, if an interruption occurs over period $(\tau_{i-1}, \tau_i]$.

8.4 Properties of Cumulative Delays

This section examines properties of the cumulative task delays by conducting computational experiments which use a set of flexible flow shop problem instances with urgent jobs as interruptions. In these experiments, no schedule revision is performed to the predictive schedule except right-shift operations to keep its feasibility when an urgent job occurs for the purpose of observing the behavior of task delays.

Table 8.2 summarizes the schemes of the computational experiments for generating problem instances against the system in Fig. 8.2. As shown in this table, we here consider five jobs families with the number of planned job equal to 1,000. It is assumed that their inter-arrival times are *iid* random variables having $\text{EXP}(\lambda')$ with arrival rate λ' regardless of their job families. Table 8.3 also shows the scheme of the experiments for urgent jobs. It is assumed that the inter-arrival times of the urgent jobs are *iid* random variables having $\text{EXP}(\lambda)$ with arrival rate λ, where $\lambda < \lambda'$.

According to the scheme in Table 8.2, we generated five predictive schedules. For each predictive schedule, we simulated, following the scheme in Table 8.3, random arrivals of urgent jobs to collect the data concerning cumulative task delay. In this simulation, we generated arrivals of the urgent jobs over period $[0, 0.7\bar{H}_1]$,

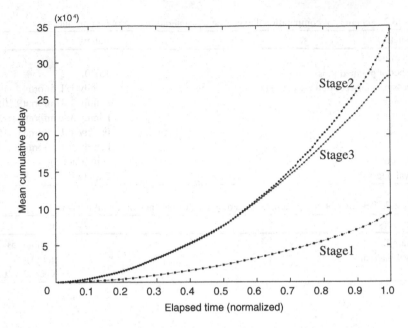

Fig. 8.3 Mean cumulative delay with $\lambda = 0.01$ observed at each stage

where $\bar{H}_1$ denotes the expected makespan of the schedule for stage 1 obtained by some preliminary experiments. The average of cumulative task delays was computed over the five predictive schedules at the time when any task was completed in each predictive schedule.

Figure 8.3 depicts the behavior of the mean cumulative delay observed at each stage. Note in this figure that the horizontal axis expresses the normalized elapsed time, $\tilde{t}$, which is defined by $\tilde{t} = (t - s_\theta)/\bar{H}_1$ ($s_\theta \leq t \leq s_\theta + \bar{H}_1$), where s_θ is the starting time of the first job at stage θ. It is observed in this figure that the mean cumulative delay at any stage increases at an accelerated rate with respect to the elapsed time. In addition, it increases more sharply at stages 2 and 3 than that at stage 1. This is because stage 2 is a bottleneck stage with a single machine and the delays at stage 2 influence upon stage 3. It is also seen that the mean cumulative delay at stage 3 is smaller than that at stage 2; stage 2 is a bottleneck stage but stage 3 consists of two machines in parallel.

8.5 Performance of the Enhanced D^*-Driven Policy

This section examines the performance of the enhanced D^*-driven policy by comparing it with the periodic revision policy through computational simulations.

Table 8.4 Simulation schemes—occurrence of urgent jobs (2)	Parameters	Values
	Job families	Uniform$(1, 5)$
	Arrival interval	Exp(λ)
		where $\lambda = 0.005, 0.01$

8.5.1 Simulation Schemes

Problem Instances

We considered 20 problem instances (manufacturing situations) in accordance with the schemes in Tables 8.2 and 8.4. In Table 8.4, we have introduced $\lambda = 0.005$ to reflect severer situations than that of $\lambda = 0.01$. The target system considered here is the same as that in Sect. 8.2.2. Under these conditions, we generated urgent jobs randomly, which were *iid* random variables having an exponential distribution with rate $\lambda = 0.005$ or 0.01.

The performance measures employed here are:

- Total completion times;
- Total setup times, and
- Frequency of schedule revision.

The planning horizon was set to $[0, 5110]$, which was empirically determined based on some basic preliminary experiments.

Design of Enhanced D^*-Driven Policy

The enhanced D^*-driven policy involves three kinds of design variables; they are D^*, T, and τ. We first focus on the critical cumulative delay, D^*. Chapter 6 examined the behavior of the mean cumulative task delay by introducing a function, $D(t)$, at time t in job shop environments and attempted to express $D(t)$ by

$$D(t) = at^b, \tag{8.4}$$

where $a > 0$ and $b \geq 1$. In this chapter as well, we consider the same structure given by Eq. (8.4) to represent the behavior of the mean cumulative task delays. For the purpose of obtaining estimates of a and b for $\lambda = 0.005$, we carried out again the preliminary simulations in the same way as in Sect. 8.4. Table 8.5 shows the results of parameter estimation for $\lambda = 0.01$ and 0.005.

Based on the estimates in Table 8.5, we have obtained various values for the cumulative task delay, D_θ^*, at stage θ by

$$D_\theta^* = a \left(\alpha \bar{H}_\theta \right)^b, \tag{8.5}$$

Table 8.5 Estimates of parameters a and b

	$a\ (\times 10^{-2})$		b	
$\lambda =$	0.005	0.01	0.005	0.01
Stage 1	1.21	2.24	1.82	1.78
Stage 2	0.18	0.07	1.92	2.08
Stage 3	0.16	0.04	1.94	2.14

Table 8.6 Critical cumulative delays

	$\lambda = 0.005$			$\lambda = 0.01$		
	Stage 1	Stage 2	Stage 3	Stage 1	Stage 2	Stage 3
α	D_1^*	D_2^*	D_3^*	D_1^*	D_2^*	D_3^*
0.10	1593	550	1090	1046	288	552
0.15	3333	1197	2393	2153	669	1315
0.20	5626	2080	4182	3593	1216	2435

where $\bar{H}_\theta$ represents the average of makespans obtained by the preliminary experiments and the value of α was empirically set to $0.10, 0.15, 0.20$.

Table 8.6 summarizes the values of critical cumulative delay, D_θ^* for $\theta = 1, 2, 3$, obtained in the above manner. In this table, D_1^* takes on a larger value than D_2^* and D_3^*, and the values of D_3^* are greater than those of D_2^*. This reflects the number of machines in each stage.

As for the other two design variables, T and τ, we use $T = 480$ and $\tau = 10$.

8.5.2 Comparison with Periodic Revision

In this subsection, we make a comparison between the enhanced D^*-driven policy and a conventional periodic schedule revision policy on the basis of the simulations results.

The previous subsection Sect. 8.5.1 introduced 20 problem instances to examine the performance of the D^*-driven policy. These 20 problem instances are used again in this subsection. Under the enhanced D^*-driven policy, we have two values of λ and three sets of design variables, (D_1^*, D_2^*, D_3^*), that are identified in terms of the value of α, and consequently we have six scenarios. Consequently, we have 120 instance-scenarios. On the other hand, the conventional periodic schedule revision policy has two design variable, the inspection time interval, τ, and the schedule revision interval, T. In this subsection, we use:

- Inspection interval: $\tau = 10$;
- Interval of schedule revision: $T = 100, 200, \dots, 1000$,

which yields a total of 10 cases. Since we consider two values for λ, the conventional policy has 20 scenarios and consequently 400 instance-scenarios.

Table 8.7 Simulation results by the *enhanced* policy

(a) $\lambda = 0.005$

α	Mean total completion time	Mean total setups	Frequency of schedule revision
0.1	514.3	1726.4	14.4
0.15	526.6	1806.5	9.1
0.2	526.3	1797.8	5.0

(b) $\lambda = 0.01$

α	Mean total completion time	Mean total setups	Frequency of schedule revision
0.1	498.7	1723.2	22.1
0.15	523.3	1816.1	12.1
0.2	532.6	1838.8	8.8

The comparison between the two schedule revision policies is carried out in reference to the mean total completion times of jobs per scenario, and the mean total setup times per scenario. We also examine the frequency of schedule revisions.

The simulation results are summarized in Tables 8.7 and 8.8 and Figs. 8.4 and 8.5. Table 8.7 shows the mean total completion times of jobs per problem instance under the enhanced D^*-driven policy. The mean total setup times and the mean frequency of schedule revisions per problem instance are also indicated in this table. Table 8.8 reveals the simulation results of the conventional periodic policy. It can demonstrably be observed in Table 8.8 that the large interval between successive schedule revisions invokes the late mean total completion times of jobs and the large mean total setup times along with the less frequent schedule revisions per problem instance, and this is because the opportunity of a schedule revision decreases with increasing interval length.

Figures 8.4 and 8.5 compare the enhanced D^*-driven policy with the conventional periodic policy from the viewpoint of Pareto optimality. The horizontal axes in both Figures reflect the mean frequency of schedule revisions per problem instance. The vertical axis in Fig. 8.4 signifies the mean total completion times of jobs per problem instance, while that in Fig. 8.5 expresses the mean total setup times per problem instance. Figure 8.4a indicates, in the case of $\lambda = 0.005$, that the enhanced D^*-driven policy notably dominates the conventional periodic schedule revision policy with respect to both the mean total completion time and the schedule revision frequency. On the other hand, Fig. 8.4b shows the results in the case of $\lambda = 0.01$, which reflects more frequent arrival of urgent jobs than $\lambda = 0.005$. We can observe in this figure that the results of the enhanced D^*-driven policy show their Pareto optimality with respect to the two criteria expressed by the horizontal axis and the vertical one. The similar tendencies to the above can be seen in Fig. 8.5a and b.

Table 8.8 Simulation results
by the periodic policy

(a) $\lambda = 0.005$

T	Mean total completion time	Mean total setups	Frequency of schedule revision
100	501.5	1740.5	43.0
200	541.8	1891.8	21.0
300	565.9	1850.0	14.0
400	569.9	1900.5	10.0
500	587.0	1936.3	8.0
600	575.1	2022.3	7.0
700	586.9	2070.0	6.0
800	583.8	2145.8	5.0
900	587.7	2194.8	4.0
1000	589.8	2275.8	4.0

(b) $\lambda = 0.01$

T	Mean total completion time	Mean total setups	Frequency of schedule revision
100	487.6	1691.4	43.0
200	509.2	1804.0	21.0
300	556.3	1825.7	14.0
400	559.7	1898.4	10.0
500	560.6	1971.2	8.0
600	560.1	2046.9	7.0
700	568.6	2054.1	6.0
800	553.0	2129.9	5.0
900	570.6	2211.2	4.0
1000	573.2	2270.8	4.0

8.5.3 Comparison with Other Cumulative Delay-Based Policies

This subsection compares the *enhanced* D^*-driven policy with the D^*-driven policy as well as the *hybrid* D^*-driven policy for the purpose of showing the advantages of the enhanced D^*-driven policy. In the following, the enhanced D^*-driven policy is called the *enhanced* policy for simplicity, whereas the D^*-driven and the hybrid D^*-driven policies are, respectively, called *standard* and *hybrid* D^*-driven policy. The criteria in this comparison are the same as those in Sect. 8.5.2.

The computational simulations were performed for the 20 problem instances which were generated according to the schemes shown in Tables 8.2 and 8.4, and were used in the previous subsection as well. The design variables of the standard policy and hybrid policy are the same as those of the enhanced policy in the previous subsection, i.e.,

- Inspection interval is $\tau = 10$;
- Critical cumulative delays are set to values as shown in Table 8.6, and;
- Interval of periodic revision is $T = 480$.

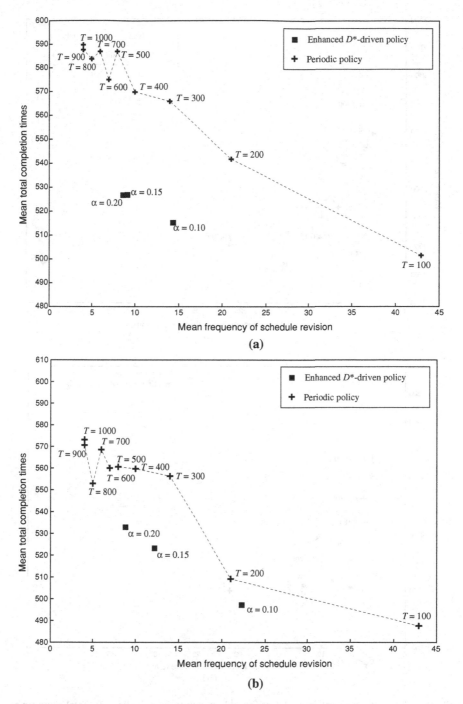

Fig. 8.4 Comparison of enhanced D^*-driven policy with periodic policy (1): Total completion times versus frequency of schedule revision. **a** $\lambda = 0.005$, **b** $\lambda = 0.01$

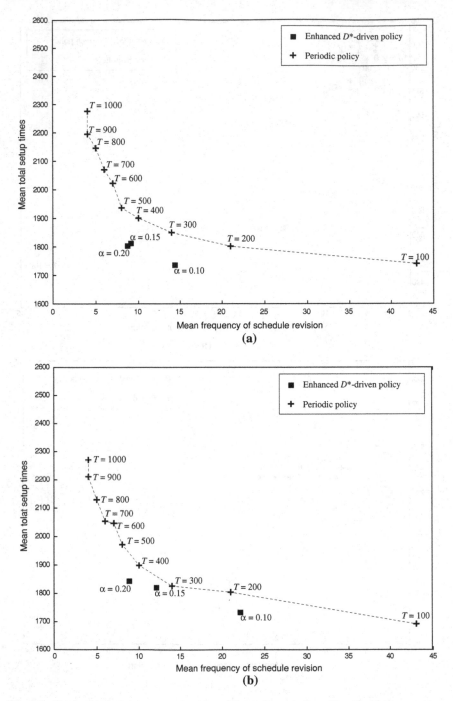

Fig. 8.5 Comparison of enhanced D^*-driven policy with periodic policy (2): Total setup times versus frequency of schedule revision. **a** $\lambda = 0.005$, **b** $\lambda = 0.01$

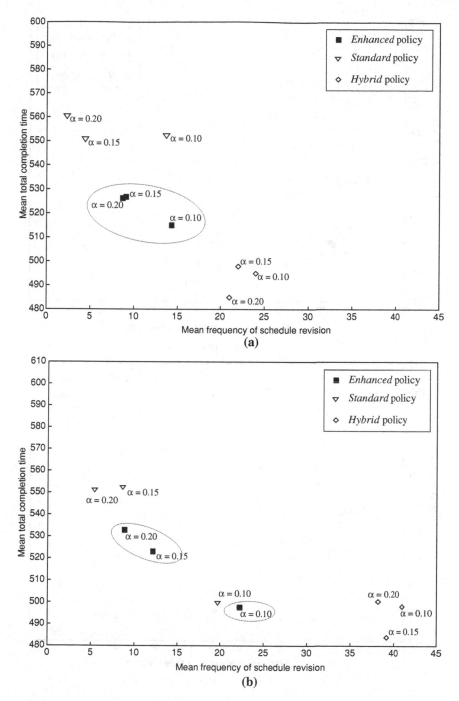

Fig. 8.6 Comparisons of D^*-driven policies (1): Total completion times versus frequency of schedule revision. **a** $\lambda = 0.005$, **b** $\lambda = 0.01$

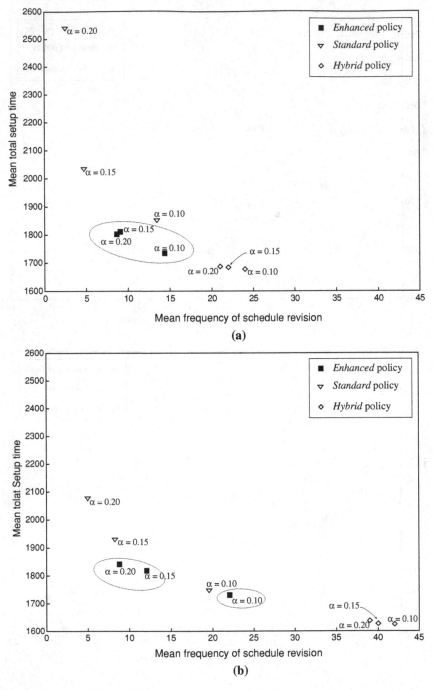

Fig. 8.7 Comparison of *D**-driven policies (2): Total setup times versus frequency of schedule revision. **a** $\lambda = 0.005$, **b** $\lambda = 0.01$

Since we consider the two cases, $\lambda = 0.005$ and 0.01, for the arival rate of urgent jobs, we have 40 instance-scenarios eventually.

Figures 8.6 and 8.7 summarize the simulation result by concentrating upon the mean total completion times of jobs, the mean setup times, and the mean frequency of schedule revisions, all of them are collected per problem instance. In Fig. 8.6a, we can observe that the enhanced policy tends to reduce the mean total completion times of jobs more effectively than the standard policy, and moreover, it tends to show less frequent schedule revisions than the hybrid policy. Figure 8.6b reveals the similar tendency but it is less distinct than in Fig. 8.6a since the former corresponds to a harsher case than the latter because of more frequent arrivals of urgent jobs.

Figure 8.7a indicates that the enhanced policy tends to show smaller values of the mean total setup times than the standard policy, but less frequency of schedule revisions than the hybrid policy. In Fig. 8.7b, furthermore, we can observe the similar tendency to that in Fig. 8.7a.

In summary, the simulation results observed in the above advocate that the enhanced policy is the most balanced among the three policies in the sense that it can reduce the total completion times of jobs as well as the total setup times with less frequent schedule revisions.

References

1. Brah SA, Hunsucker JL (1991) Branch and bound algorithm for the flow shop with multiple processors. Eur J Oper Res 51:88–99
2. Krüger K, Sotskov YN, Werner F (1998) Heuristics for generalized shop scheduling problems based on decompositions. Int J Prod Res 36(11):3013–3033
3. Lee HF, Stecke KE (1998) Production planning for flexible flow systems with limited machine flexibility. IIE Trans 30:669–684
4. Ovacik IM, Uzsoy R (1997) Decomposition methods for complex factory scheduling problems. Kluwer Academic Publishers, Massachusetts
5. Pflughoeft KA, Hutchinson GK, Nazareth DL (1996) Intelligent decision support for flexible manufacturing: design and implementation of a knowledge-based simulator. Omega, Int J Manage Sci 24(3):347–360
6. Pinedo M (2008) Scheduling—theory, algorithms, and systems, 3rd edn. Springer, New York
7. Quadt D, Kuhn H (2007) A taxonomy of flexible flow line scheduling procedures. Eur J Oper Res 178:686–698
8. Sriskandarajah C, Sethi SP (1989) Scheduling algorithms for flexible flowshops: worst and average case performance. Eur J Oper Res 43:143–160
9. Suwa H, Sandoh H (2007) Capability of cumulative delay based reactive scheduling for job shops with machine breakdowns. Comput Ind Eng 53:63–78
10. Vieria GE, Herrmann JW, Lin E (2000) Predicting the performance of rescheduling strategies for parallel machine systems. J Manuf Syst 19(4):256–266
11. Wittrock RJ (1985) Scheduling algorithms for flexible flow lines. IBM J Res Dev 29(4):401–412
12. Wittrock RJ (1988) An adaptable scheduling algorithm for flexible flow lines. Oper Res 36(3):445–453

Index

H. Suwa and H. Sandoh, *Online Scheduling in Manufacturing*,
DOI: 10.1007/978-1-4471-4561-5, © Springer-Verlag London 2013